ADG-4300

Cenyp
~~SY~~

Science Concepts SECOND SERIES

Growth and Development

Alvin Silverstein, Virginia Silverstein, and Laura Silverstein Nunn

Twenty-First Century Books
Minneapolis

Twenty-First Century Books
A division of Lerner Publishing Group, Inc.
241 First Avenue North
Minneapolis, MN 55401 U.S.A.

Website address: www.lernerbooks.com

Library of Congress Cataloging-in-Publication Data

Silverstein, Alvin.
Growth and development / by Alvin Silverstein, Virginia Silverstein, and Laura Silverstein Nunn.
p. cm. — (Science concepts)
Includes bibliographical references and index.
ISBN-13: 978–0–8225–6057–9 (lib. bdg. : alk. paper)
1. Growth—Juvenile literature. 2. Developmental biology—Juvenile literature. I. Silverstein, Virginia. II. Nunn, Laura Silverstein. III. Title.
QH511.S53 2008
571.8—dc22 2006030299

Manufactured in the United States of America
1 2 3 4 5 6 – DP – 13 12 11 10 09 08

Contents

chapter one

Life Stories

Have you ever seen a caterpillar "magically" change into a butterfly? Have you ever planted a young tree in your backyard and then, as time went by, noticed that it was much bigger than you remembered it? Did you ever have a cute little puppy and watch it grow into a big, strong dog in just six months? Have you ever looked in old family photos and seen how your parents changed through the years—from tiny babies to children, teenagers, and then adults? All living things—from insects to trees to animals to people—have their own life story. Each one goes through a life cycle in which great changes occur through growth and development.

At some point in their life cycle, living things may reproduce—have babies that will go through the same process of growth and development as their parents. They form a population that can grow and develop too.

These sixteen-day-old golden retriever puppies cuddle close to their mother to nurse. These puppies will change quickly in the coming months and be full-grown dogs at one year old.

Life Changes

From the moment you are born until the day you die, your body continually goes through changes. These changes happen over many, many years. However, the most dramatic changes in your life happened over a very short period of time, and they actually occurred *before* you were born.

You started out as a single cell. By the time you were born, your body contained trillions of cells. In the nine months from when you were conceived until you were born, your body increased about ten billion times in size.

Growth vs. Development

Many people think that *growth* and *development* mean the same thing. Actually, though, growth simply refers to getting bigger. The development of living things is a process. It may include growth, but it also includes a series of changes in shape or form. For example, as a Labrador puppy grows into a dog, it gets much larger. But it also changes. Its face gets longer, changing from the rounded puppy shape to the pointed muzzle of the dog. Its legs get longer too, and its body grows much faster than its head. Development also includes internal changes that cause the puppy to get stronger, better coordinated, and smarter as it grows into a dog.

After babies are born, they still grow fairly rapidly. During their first year, babies usually triple their weight and add about 50 percent in height. In the following years, however, children grow more slowly. Most kids gain an average of 5 to 6 pounds (2.3 to 2.7 kilograms) and grow 2 to 3 inches (5.1 to 7.6 centimeters) a year. This changes when they become teenagers. During adolescence most kids have a sudden growth spurt, growing as much as 7 inches (17 cm) in a year. After that, growth slows down again, and once they are fully mature, growth will stop entirely. Changes in their body and mind will continue, though, for the rest of their lives.

Different Patterns

Many animals have a pattern of development similar to humans, but the lengths of their lives differ. The amount of time they spend in the different stages of life—childhood, adolescence, and adulthood—also differs, depending on the species. Kittens and puppies, for example, are able to have babies of their own in just six months and are fully grown within a year or two. Few of them live longer than twenty years or so. Mice are fully grown and can reproduce just a few weeks after they are born, and they usually live for less than two years. Elephants, like humans, are mature by about thirteen years of age and can live to about seventy. Humans can live for as many as one hundred years, but on average, people can expect to live into their seventies or eighties.

If you think the changes that transform a baby into an adult are remarkable, you will be amazed at the kinds of changes some animals go through during their

Did You Know?

Fish and tortoises go on growing throughout their lives, but the growth can be very slow. Scientists are not sure exactly how long they can live, but some of the Galápagos tortoises that the famous biologist Charles Darwin observed back in 1835 are still alive!

life cycle. A baby frog is a tadpole, which looks nothing like an adult frog. It looks more like a fish and spends all its time in the water. To become a frog, it will lose its tail and grow front and hind legs. Other major changes will also occur in its body shape, its eyes and mouth, and its systems for breathing and getting food.

Although these changes are dramatic, they occur gradually. The changes that transform a caterpillar into a butterfly are even more fascinating. A wiggly, wormlike creature with tiny eyes and a lot of stubby little legs turns into a hard-bodied insect with six long legs, two "feelers" on its head, and two spreading pairs of wings that can lift it into the air. What's more, the change is not gradual. When a caterpillar is ready to turn into a butterfly, it encloses itself into a hard shell and does not come out until it is completely changed.

Plants have their own patterns of growth and development. They, too, start out as a single cell. How they get from there to the mature adult depends on the kind of plant. Flowering plants, for example, start out as seeds. Seeds may be hard and dry, looking no more alive than a pebble or a grain of sand. Give them some warmth and moisture, though, and a tiny shoot soon pokes out and grows into a seedling. This tiny plant grows and develops, forming new leaves, stems, and roots. Eventually it produces seeds that can grow into the next generation.

Some plants continue to grow steadily for years.

Others have periods when growth stops for a while during unfavorable weather and other conditions—throughout the cold winter or the hot, dry summer, for example. Still others live for a single season, which may last for months or weeks or even just a few days—long enough to produce seeds that will start a new generation the following year. The chapters that follow will take a closer look at the many patterns of growth and development in the living world.

This seed has sprouted out of the ground and is growing and developing into an adult pea plant.

chapter two

How Do Things Grow?

When you feel hungry, do you get a snack, such as an apple or a granola bar? Feeling hungry is your body's signal that you need to take in more raw materials. These materials are used for two main things: energy to power your activities and building materials for growth and repair.

You are still actively growing, so you probably feel hungry very often. But adults, who are no longer growing, need to eat too. Each day many body cells get worn out or damaged and have to be replaced. So people of all ages need to continue "fueling up" to keep their bodies in working order.

An apple or a granola bar cannot be used just as it is to build new parts for a growing person. Its materials must first be broken down into simpler chemicals and then changed to make them into *human* materials, such as tissues or muscles. All living things take in materials from outside and rebuild them into materials they can use.

Life Cycle of a Cell

Throughout your life, from the first cell to a baby, a

Body cells come in all shapes and sizes. This artificially colored microscopic image shows various red and white blood cells.

child, and, finally, an adult, you grow as the number of cells in your body increases. This happens by cell division, a process in which a single cell grows and splits into two cells. This kind of cell division is called mitosis.

Mitosis is necessary for all growth, whether it is a tree growing longer limbs or a child growing taller. As an organism grows, mitosis is responsible for building and developing until the organism reaches its full potential.

What Are Cells?

Cells are the basic units of life. The trillions of cells that make up your body have many different shapes—round or flat, tiny squared-off cubes, or jellylike blobs. (You need a microscope to see them.) Your body cells are specialized for different jobs, but they work together to keep your body going. Although cells are alive, none of your body cells could survive on their own. Some very simple organisms, though, are made up of just a single cell. Each one can do many of the same things you do—move, eat, detect light, and react to its surroundings.

Mitosis

1. Prophase

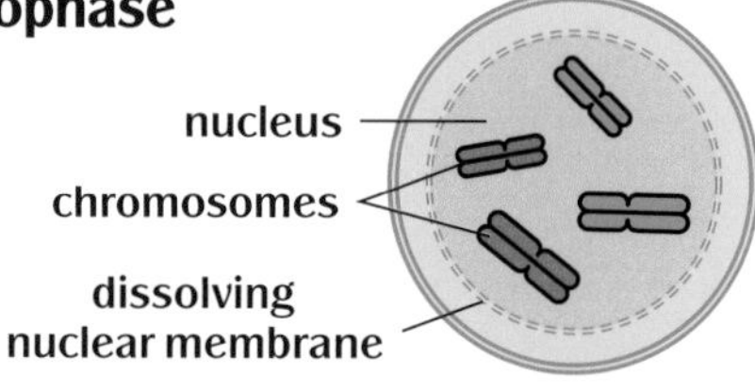

Previously copied chromosomes coil up and become visible. The nuclear membrane, a double membrane surrounding the nucleus, dissolves.

2. Metaphase

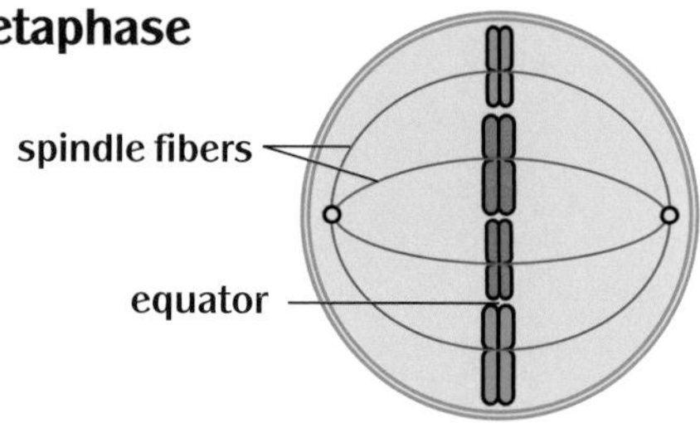

The chromosomes move to the center of the cell and line up along the "equator." Spindle fibers link each half of a chromosome to opposite "poles."

3. Anaphase

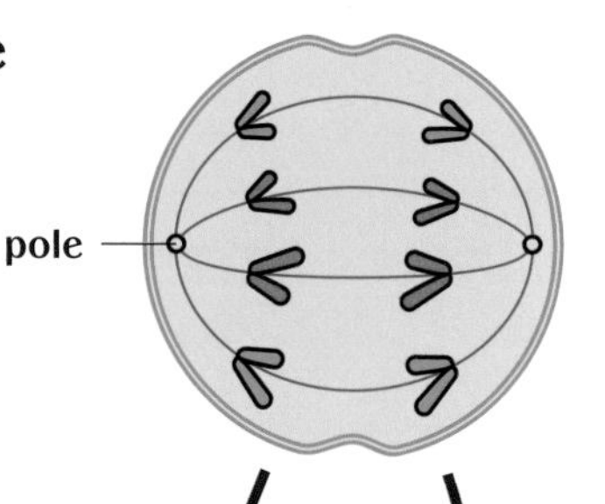

The chromosomes split, and the two halves travel to opposite poles as the spindle fibers shorten.

4. Telophase

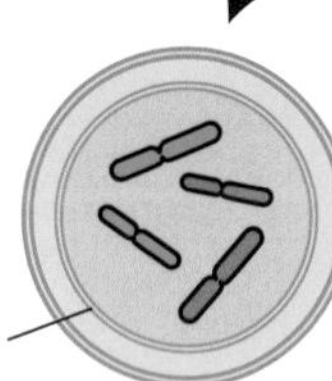

New nuclear membranes form around the chromosomes of the two daughter cells produced. The spindle fibers dissolve.

But mitosis doesn't stop there. Cells that are damaged, diseased, or worn out have to be replaced. Every second, millions of cells in your body die. If you couldn't replace these lost cells, you would die

> **Did You Know?**
> You lose as many as forty thousand dead skin cells every time you wash your hands!

in a matter of days. It is through mitosis that these dead cells are replaced and you can go on living.

Normally, cells reproduce at just the right rate to replace the dying cells. But how do dividing cells know when to stop? Cell division is generally controlled through chemical signals sent by certain cells to other cells in the body. Contact with the cells around it may also tell a body cell to divide or not to divide.

Before mitosis can begin, a cell must first make a copy of its DNA (deoxyribonucleic acid). This is a chemical that contains instructions for all the cell's work and blueprints for making new cells. Genes, the units of heredity, are made of DNA. Nearly every living cell contains a complete set of genes.

A cell that divides is called a mother cell, and its offspring are known as daughter cells. When a cell divides by mitosis, the two daughter cells that are formed are smaller copies of the original cell. Each daughter cell gets a complete copy of all the DNA instructions its mother had. All the other structures and materials in the original cell are also divided between the two daughter cells. When the daughter cells become large enough to divide, then they can become mother cells to their own daughter cells, passing along the same genes they inherited from their mother . . . and the process continues.

Mitosis takes place when bacteria reproduce. Bacteria are made up of just one cell. When the cell gets to a certain size, it makes a copy of all its DNA. Then parts of its outer covering grow down the middle of the cell, separating the old and new

Did You Know?

Under the right conditions—warmth, moisture, and plenty of food—bacteria can multiply as often as once every twenty minutes. Their numbers can increase at a really fast rate!

sets of DNA. The cell divides into two daughter cells. These daughter cells separate and go off to lead their own individual lives. With a complete set of genes and just the right building blocks, the daughter cells can grow, and the cycle starts all over again. This process is known as binary fission (which means "splitting into two").

In multicellular organisms, which are made up of many cells, mitosis occurs in much the same way as it does in bacteria. But the daughter cells that are formed do not usually separate. They stay linked, adding to the growth of the whole organism. Blood cells and reproductive cells (sperm and egg cells) are exceptions. They separate after dividing, much like single-celled organisms.

Individual Timetables

When cells in a multicellular organism divide by mitosis, the daughter cells may be exact miniature copies of their mother cell or they may be quite different. How can that be, when all of them have the same set of genes? The answer is that most of a living cell's genes are normally covered with proteins that keep them from working actively. In fact, most of the genes in a cell are actually "turned off" most of the time. When an organism grows and its cells divide, certain genes are uncovered and "turned on" in one kind of cell, while different genes are turned on in other kinds of cells. In each cell, different genes are turned on or off, depending on the cell's needs.

The cell's "book of instructions"—its DNA—includes a timetable for the development of the organism. As the original cell multiplies into an organism, genes determine what new parts are needed, where they will form, and when they will be produced. In addition to the built-in timetable, various signals from outside the cell—chemical messages,

Sizing Up Cells

Cells come in all sorts of shapes and sizes, and they vary greatly. The smallest cells in the world are mycoplasmas, germs that cause pneumonia in humans. They measure only 150 to 200 nanometers in diameter. If about seven thousand of them were lined up in a row, they would measure only 0.04 inch (1 millimeter). The largest cell is the yolk of an ostrich egg, which is about the size of a baseball. The longest cells are nerve cells that run down the neck of a giraffe. They may be more than 9.7 feet (3 meters) long. But even these giraffe nerve cells cannot be seen without a microscope. That's because nerve cells are extremely thin—thinner than the hairs on your head and even thinner than the strands of a spiderweb.

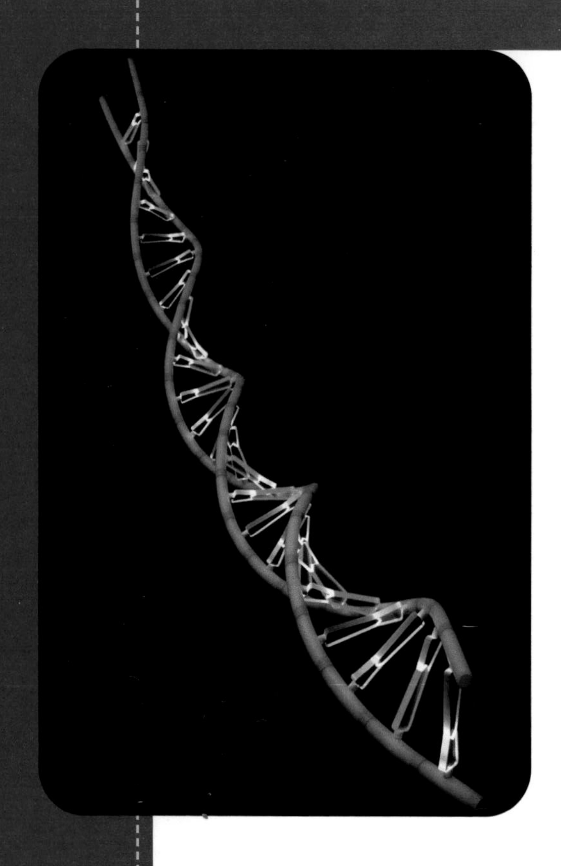

This model shows the structure of DNA, which looks like a spiral staircase. This structure is called a double helix.

heat, light, gravity, and others—help to determine which genes should be working.

Genes direct the production of proteins. As each cell makes its own particular kinds of proteins, it becomes different from the other cells in the organism. With each successive cell division, the various kinds of cells become more and more different. This process is known as differentiation.

Cells differentiate to form the various parts of the body, each with its own function. The specialized cells group to form body tissues. A grouping of tissues in a specific structure forms an organ, such as a heart or kidney in an animal or a leaf or stamen in plants.

Growth Gone Wild

Many people think that cancer is the scariest disease of all. In cancer the body's own cells turn against it and run wild. Unlike normal cells, cancer cells don't know when to stop growing. This happens because, at some point, the cancer cells mutated (changed) in various ways and lost the ability to pick up and respond to signals that control cell division. So they just keep dividing. They crawl over other cells and push their way into healthy tissues. They may form a large mass called a tumor.

Cancer cells lose some of their specialized functions and do not work for the good of the body. Soon they start to choke out normal, healthy cells that have important jobs to do, such as making blood, digesting food, or controlling the movements of body parts. The cancer cells also steal food and other materials the body cells need to live.

Cancer cells can cause so much damage to the body cells that they may kill the organism. Cancer can affect people at any age. Cancerous tumors can also develop in most kinds of animals and even many plants.

Did You Know?

The largest tumor ever removed from a person weighed 303 pounds (137 kg)! It was a mass of tissue in a woman's abdomen, and it measured 3 feet (1 m) in diameter. After the operation, the patient weighed 210 pounds (95 kg)—down from more than 500 pounds (227 kg) before the operation.

Researchers hope someday to find ways of using genetic on-off switches to make cancer cells stop growing wildly and differentiate into normal tissues.

Not all tumors are cancerous. Sometimes cells lose only part of their ability to respond to the controls on cell division. They may multiply wildly, forming huge masses, but they do not invade the cells around them. Doctors call such tumors benign. Cancerous tumors are referred to as malignant.

Actually, most tumors are benign. They range from warts (small tumors on the skin, formed when a virus infection causes skin cells to overgrow) to tumors as big as a basketball or even larger.

This plant has crown gall disease, which causes soft tumor-like growths called galls to form on the plant.

Monster Growths

Usually cancerous tumors have lost much or all of their differentiation and are just blobs of tissue. But tumors called teratomas are very strange. Their name comes from a Greek word meaning "monster." Teratomas contain several different kinds of tissues. Although they are not organized into a working organ, they may contain a mixture of organ parts—including hair, fingernails, and even teeth!

Depending on where they grow, benign tumors may be just a nuisance or they may become dangerous. Benign brain tumors can cause headaches, loss of speech or other abilities, and even death because the hard, bony skull cannot expand. As the tumor grows, it squashes working brain cells and may cut off their blood supply.

chapter three

Growing Up Human

Does your mother or father keep a record of how much you grow from year to year? Maybe there is a special wall in your house that has pencil marks of your height at different levels, and next to each one is the date and height measurement. If you take a look at your homemade growth chart, you can see how much you have grown over the years, inch by inch.

You might also have pictures of you taken at various ages. Looking at them you can see a lot of changes—in your height and weight, and in your hair, face, and body. The height chart and all the pictures, however, do not include a very important part of your life—the nine months or so before you were born.

Before You Were Born

Before birth every human being is a water animal. For nine months a human baby grows inside its mother's uterus, or womb—a pear-shaped organ in her belly. The baby floats in a liquid-filled chamber very much like a space capsule, which protects the baby from bumps and shocks.

Inside your mother's womb, you started out as a single cell called a zygote. This is a fertilized egg that was formed when a sperm (male sex cell) from your father and an egg (female sex cell) from your mother joined inside one of your mother's two fallopian tubes.

For a short time after fertilization, nothing much seems to happen. But then, suddenly, after about thirty hours, the single cell of the zygote splits in two and becomes a developing embryo. About twenty hours later, the two cells split again to form four cells. The splitting occurs again, and eight cells are formed, then sixteen and thirty-two and sixty-four. So far there has not been any growth—the many-celled embryo is no bigger than the original zygote.

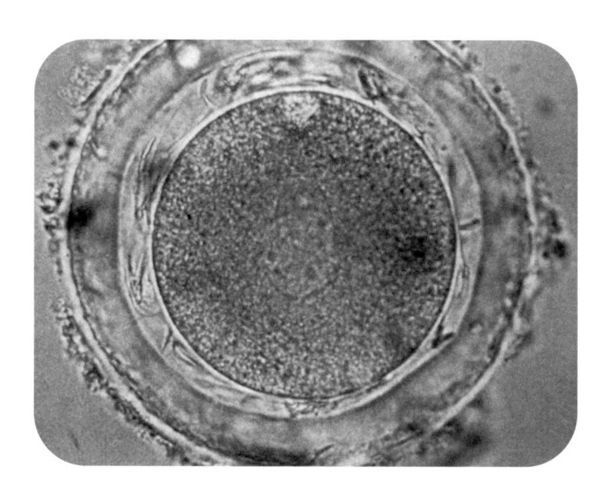

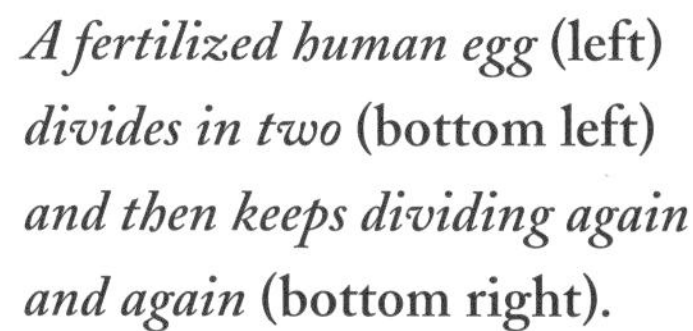

A fertilized human egg (left) *divides in two* (bottom left) *and then keeps dividing again and again* (bottom right).

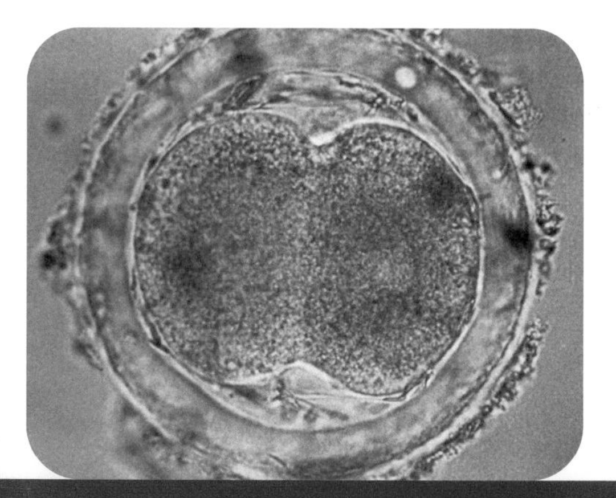

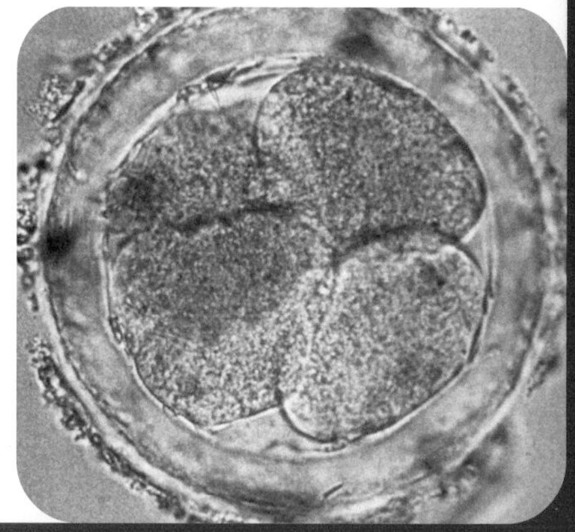

As the cells divide, they stay clustered like berries. By the time the embryo has reached the uterus, it has become a fluid-filled, hollow ball of cells. About a week after fertilization occurs, the little embryo settles down in the lining of the uterus and buries itself into the rich, nourishing tissue. A net of fingerlike blood vessels fans out into the mother's tissues and soaks up nourishment for the growing embryo. This will develop into the placenta, a spongy structure that will bring nutrition from the mother to the embryo. The placenta thus grows from the embryo's tissues, but it becomes closely linked with the mother's tissues as it grows into the lining of the uterus. The placenta makes chemicals that help the mother's body undergo the changes necessary to accommodate the new life growing within her.

Did You Know?

For the first twenty years of life, as a child grows from an infant to an adult, the child's height can increase about three and a half times and weight more than twenty times. But in just the first eight short weeks inside the mother's womb, the increase is about 240 times in length and one million times in weight.

About ten to twelve days after the sperm and egg join, some cells from the embryo form the amniotic sac. About two or three weeks later, liquid (amniotic fluid) begins to build up inside it. From then on, the amniotic sac will be a kind of space capsule, in which the developing embryo floats, protected from harm. The embryo is attached to the wall of the capsule by a thick, ropelike tube, called the umbilical cord. As the placenta forms, blood vessels develop inside the umbilical cord. They carry blood back and forth between the embryo and the placenta.

Meanwhile, during the second week after fertilization, some of the cells on the outside of the hollow ball move inward, forming first two and then three layers. The growing embryo is then disk-shaped. At first its cells are identical to one another. But then cells of different shapes, sizes, and functions appear. The cells of the outer layer will form the skin, the brain, and the nerves. Those in the inner layer will become the lining of the digestive system and internal organs. The cells

When a human being begins to develop, the fertilized egg (zygote) divides again and again. It forms a berrylike cluster of cells (morula) and then a hollow ball (blastula). As the cells continue to divide, they differentiate to form the gastrula, with first two and then three layers of cells. All the organs of the body develop from these first cell layers.

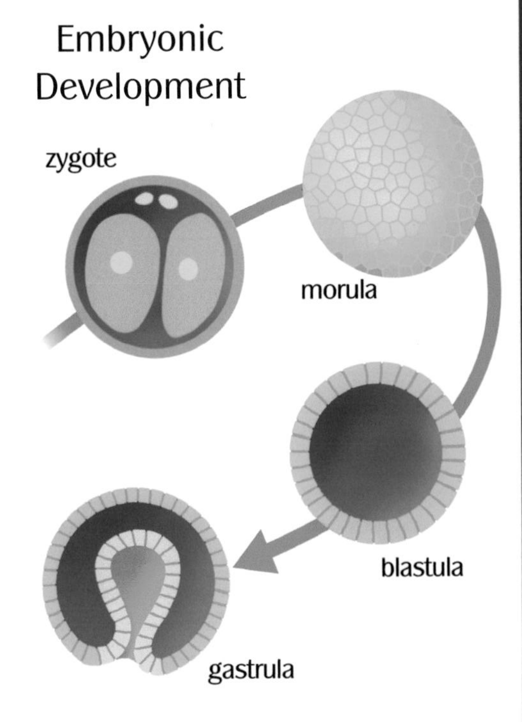

of the middle layer will develop into muscles, bones, and the heart and blood vessels. A long groove appears on the embryo, then pinches together and joins to form a tube. This will be the spinal cord and brain. A bulge grows at one end and bends over. This will be the head. Another bulge appears below the head. This will be the heart, and it begins to beat at about four weeks. Meanwhile, bands of tissue that will become muscles and bones have been forming, and various organs are taking shape.

At four weeks it is hard to tell what the little creature will be like. It has a long tail and gill slits like a fish. It is only 0.25 inch (0.6 cm) long, but it is ten thousand times heavier than it was when it was just one cell.

At about five weeks after fertilization, little buds appear, which will become the arms and legs. The eyes are starting to develop, and the heart is beating more strongly inside the plump little

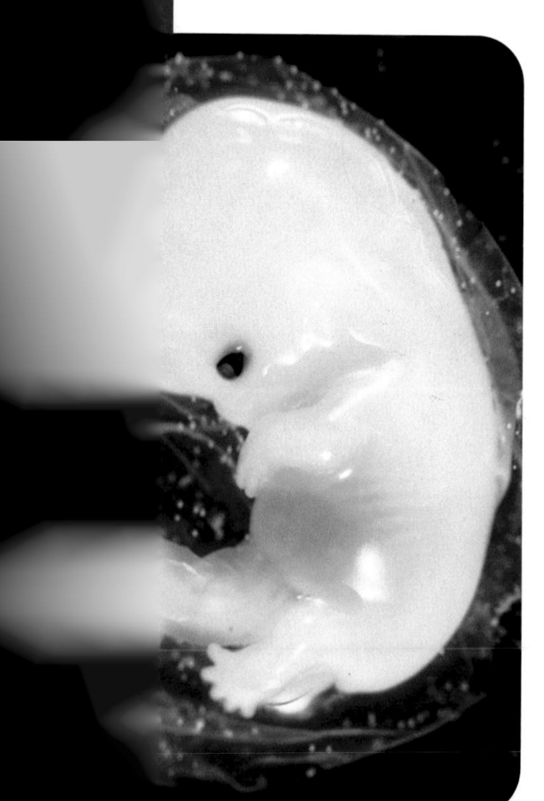

This four-week-old embryo is shown in its amniotic sac. At this stage, the embryo is about the width of a pencil eraser.

belly. At this age, there is very little difference between a human embryo and that of a donkey, rabbit, whale, bat, or lizard. But by the seventh week, the embryo is starting to look human.

The embryo continues to grow. Fingers and toes are forming, and the body grows longer and covers the tail. About two months after the zygote was formed, the embryo is just about 1 inch (2.5 cm) long. You could not tell by looking at it whether it is going to be a boy or a girl because the sex organs have not fully developed yet. But the embryo's genes have already determined which sex it will be.

By the ninth week, the embryo is called a fetus. Just big enough to fit into a peanut shell, it has all the parts that are found in a human being. It has visible sex organs, fingerprints, and eyelids that close over its eyes. The proportion of body parts, however, is very different than it will be when the baby is born—the head is about one-half the length of the fetus. (An adult's head is only about one-eighth the length of the body.)

The fetus swallows and breathes in amniotic fluid, but it does not drown. It gets the oxygen it needs from its mother's body through the umbilical cord, which attaches the fetus's belly to the placenta in the lining of the mother's uterus.

From the beginning of the ninth week to the end of the twelfth week, the fetus doubles its size—to about 2.8 inches (7 cm) long. During the third month, the trunk and limbs grow more quickly. Bone begins to form in the body. By the end of the third month, the fetus still weighs less than 1 ounce (28 grams), but it can swallow, frown, and kick its legs.

A developing fetus's growth is often measured in trimesters—that is, the pregnancy is divided into three parts,

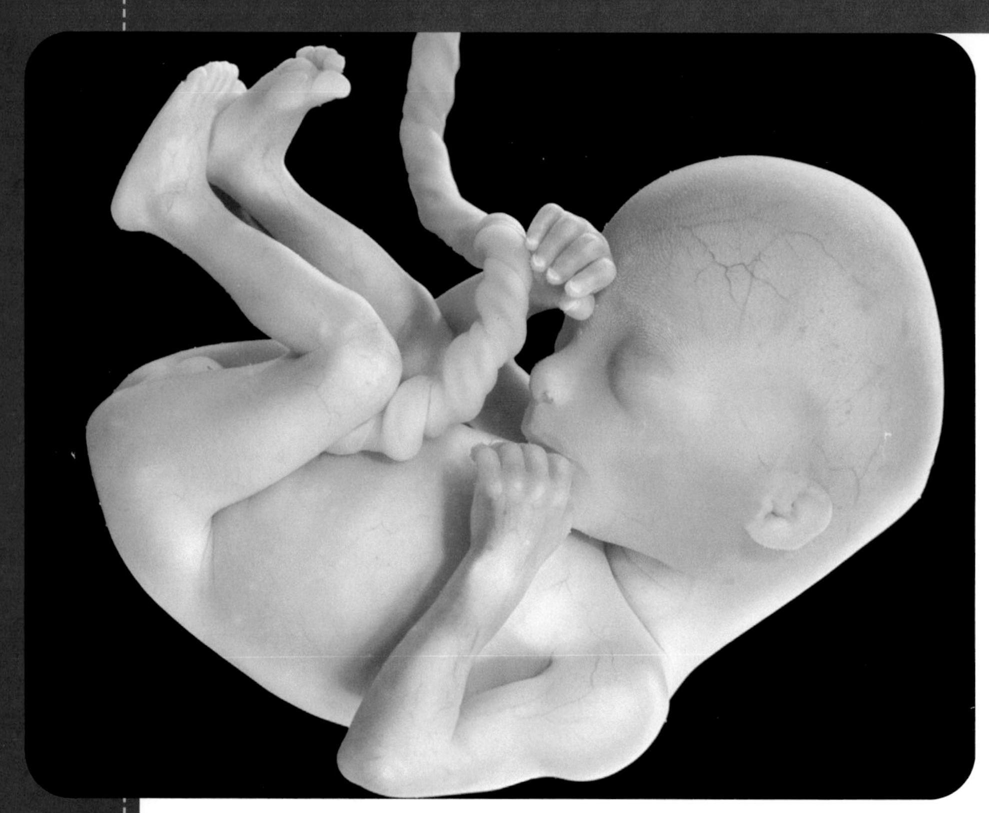

This fetus—shown with its umbilical cord—is sixteen weeks old. It's at the beginning of the second trimester of pregnancy.

each three months long. The second trimester begins in the fourth month. By then the fetus looks much more human. Its eyes, ears, nose, and mouth are well formed. Hair begins to grow on its head, and eyelashes and eyebrows appear. The heartbeat can be heard by a doctor with a stethoscope. By the end of the fourth month, the fetus is about 9 inches (23 cm) long, and the head is only one-third of the fetus's total length. The fetus can suck, drink, and get rid of some of its wastes as urine and feces. (They pass into the amniotic fluid and are carried to the placenta through the umbilical cord. The fetus's wastes pass into the mother's bloodstream and are eventually eliminated with her own urine and feces.)

In the fifth month, the entire body of the fetus becomes covered with a coat of fine hair. The skin is also covered with a waxy coating that keeps it from getting soggy. The fetus's face is red and wrinkled, and its eyelids open and close. After five months of development, the fetus has grown to about 12 inches (30 cm) long and weighs more than 1 pound (0.5 kg).

By the end of the sixth month, the fetus has completed the second trimester of its development. It is about 14 inches (35 cm) long and weighs 2.5 pounds (1 kg). It may have a full head of hair, and the fine body hair has largely disappeared. The fetus may frequently suck its thumb.

The final three months of pregnancy are a "finishing off" period. Fat deposits form beneath the skin. The fetus sleeps and wakes up, kicks, startles, and sucks its thumb. It may turn many times before settling into a head-down position, the way babies are usually delivered. At the end of the ninth month, about 266 days after the zygote began its development, the fetus weighs about 7 pounds (3 kg), measures 20 inches (51 cm) long, and is ready to be born. The baby has grown from a single cell to more than 6 trillion cells.

Did You Know?

The place where the umbilical cord attaches to the belly will become the belly button, or navel, after the baby is born.

Welcome to the World

In a very short time, the baby's body must make many important changes. For nine months, the baby has been an underwater animal, constantly wet and warm, and has received food and oxygen through an umbilical cord. As soon as the baby is born, the doctor quickly cleans fluid out of the nose and mouth. The cold air causes the baby to draw in a gasping breath, and the lungs begin to work for the first time as the baby begins to cry. (Sometimes the doctor has to help the baby start to breathe by slapping the baby's bottom.)

> **Did You Know?**
>
> As a fetus grows, so does the placenta. By the time the baby is born, the placenta has grown to about 1 inch (2.5 cm) thick and 8 inches (20 cm) long.

The baby's heart begins to work differently. Before birth a hole in the muscular wall separated one side of its heart from the other, and blood from the two sides mixed freely. This hole then closes, and the blood from the two sides of the heart no longer mixes. The blood vessels that delivered blood in and out through the umbilical cord close too. The blood that was sent to the placenta before birth now goes to the lungs instead. The baby will get new supplies of oxygen from the air, not from the mother's body.

Meanwhile, the doctor ties and cuts the umbilical cord. Now the baby is not getting any

This newborn baby boy has a typical reaction to the harsh environment outside his mother's womb.

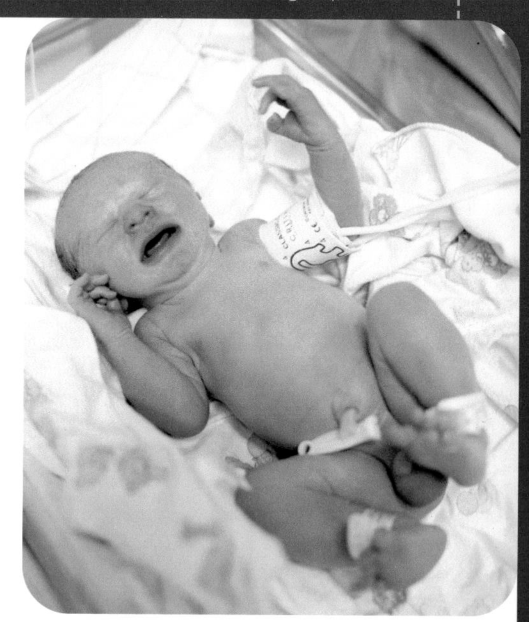

more nourishment from the placenta. The baby will have to eat and get rid of waste products.

The air is colder than the fluid inside the mother's womb, and the baby's body must learn to keep a steady body temperature. The baby cries for a bit and then blinks at the bright lights and gets a first look at the new world.

Life after Birth

A human being continues to grow physically and intellectually for about twenty years after birth. Then physical growth stops, but learning continues throughout life.

Life after birth is often divided into five stages, although when these stages occur varies quite a bit:

1) The newborn period lasts from birth through the first few weeks of life. During this period the body has to adapt to a completely new environment.

2) Infancy begins at the end of the newborn period and continues through the end of the first year, when the infant can stand alone and walk. A baby's first year is filled with major events. Gradually the baby begins to respond to the world, first to the mother's touch and voice, and later to other humans and objects. A newborn has very little coordination. Steadily, the infant gains bodily control, learning to focus the eyes, to reach and grasp with the hands, to turn over, to sit, to crawl, to stand, and finally to walk. The first baby teeth appear, usually around six months, and the baby begins to eat foods in addition to drinking milk. The baby learns to communicate, first by crying, then babbling, and finally saying recognizable words. Meanwhile, there is a steady physical growth too. By the end of the first year, a baby's birth weight will triple.

Did You Know?

You have more bones than your mother or father. Each of your parents has 206 bones, but you have dozens more. Bones do not disappear as you grow older—some of them grow together. A newborn baby has more than 300 bones!

3) Childhood lasts from the end of infancy to puberty (the time of sexual maturation), which begins at about ten years of age in girls and about twelve in boys. Children grow steadily during this period at a

rate of about 5 to 6 pounds (2.3 to 2.7 kg) a year. Children usually get thinner and stronger as they grow taller. Meanwhile, the baby teeth begin to fall out, and the first permanent teeth come in, at about six or seven years old.

Bones for a Lifetime

While you were still developing inside your mother's body, you already had a complete skeleton. That skeleton was not made of bone, though. It was made of soft, rubbery cartilage. Little by little, the cartilage was replaced by bone. Even after you were born, part of your skeleton was still cartilage. As you grew, bone continued to fill in the cartilage "model" of your skeleton. The hard bone added support, so you were soon able to hold your head up, to sit, to stand, and to walk. By the time you are twenty-five years old, your bones will be hard and strong, but the cartilage in your nose and ears will remain as cartilage for your entire life.

During childhood, children depend on their parents to provide necessities, but slowly they learn the skills they will need to make it on their own when they have matured. By two years old, a child is very active and copies many things his or her parents do. Two-year-olds learn by trial and error and imitation. They can't think things out for themselves yet. At three the child can do many things—from building a tower with blocks to riding a tricycle and working simple jigsaw puzzles. A three-year-old is also curious, constantly asking questions.

Four-year-olds can do many things for themselves, such as eating and dressing. They know how they are expected to behave. During the elementary school years, children learn more about the world. People continue to learn throughout their lives, but children learn more quickly and easily than adults.

4) Adolescence begins with sexual maturation and extends through the teen years, until about the

This group of adolescents, ages twelve to fourteen years, waits in line for lunch. The difference in height is typical of girls and boys at this age.

Growing Up

Girls usually have their growth spurt about two years earlier than boys. That's why many girls are taller than boys their age for a couple of years. Eventually, the boys will catch up—and then they grow taller than the girls. Most girls reach their full height in their mid-teens. Boys usually reach their full height in their late teens.

age of nineteen in girls and twenty-one in boys. Puberty occurs at the beginning of adolescence. Many adolescents go through a growth spurt, when they may grow several inches in a short time.

During puberty, the sex organs mature and become capable of reproduction. Chemicals called hormones are released in the body, causing various changes. For example, pubic hair (body hair around the sex organs) appears in both sexes, the breasts grow, and the sex organs grow in size. A boy's sex organs begin to produce sperm. A girl's egg cells begin to mature at the rate of one each month. Although adolescent boys and girls are not fully grown, they can already produce children of their own.

Males' voices become deeper at puberty, and their shoulders get broader. They have more body hair than females at this age and also begin to grow facial hair. Boys grow taller, and their muscles become bigger and stronger. In females the pelvis gets wider and the layer of fat under the skin becomes thicker, making females look softer.

5) Adulthood extends from the completion of adolescence until old age. The body is fully mature by early adulthood.

Aging begins in the twenties and continues to cause changes in the body throughout adulthood.

A gradual decline in muscle mass and strength occurs. Healing also starts to slow down. Bones may be weakened by a loss of calcium salts, and joints become stiff and painful as the cushioning cartilage on the ends of the bones thins and breaks down. Organs begin to work less effectively.

The body's defenses also grow less effective as the years go by, and older people may have a harder time fighting off infections. Colds may be more severe than they used to be, and they may last longer.

Digestion may be affected as well. Many older people have strict diets because their bodies cannot handle certain foods the way they used to. A common problem is lactose intolerance, the inability to digest milk and other dairy products. This problem may cause an upset stomach or diarrhea.

What Controls Growth?

To a great extent, hormones control how much we grow and when we stop growing. One of the most important of these is human growth hormone (HGH). It is produced in the pituitary gland, a pea-sized structure just under the brain. A person whose pituitary gland does not make enough growth hormone may be very short or even a dwarf. If too much is produced, a person may become a giant.

Robert Wadlow (1918–1940) would tower above a 7-foot (213 cm) basketball player. He still holds the Guinness world record as the world's tallest man. Wadlow, shown here being measured for a jacket, stood 8 feet 11 inches (272 cm) tall and wore size 37 shoes.

Basketball players who are more than 7 feet (213 cm) tall probably had an unusually high amount of growth hormone in their blood during their growing years. Most people's pituitary glands produce just enough human growth hormone so that they can grow to a height of between 5 and 6 feet (152 and 182 cm).

The food we eat also helps to determine how tall we will grow. Eating enough protein is especially important during the growing years because proteins are one of the basic building blocks for new tissues. Vitamins and minerals are important too.

As you grow taller, your bones are growing larger. You must have new supplies of calcium and phosphorus each day to

Nutrition Is Important

In the United States, boys and girls are maturing earlier than those of past generations. Scientists believe that better nutrition may be the main reason for the trend. In developing countries, where malnutrition is more common, many children are unable to grow and develop properly. They have thin and weak bones. Their organs may become underdeveloped, as well. Often these children become so weak that they are unable to fight off infections effectively. The children of immigrants from developing countries who live in developed countries, such as the United States, typically grow taller than their parents or grandparents.

build strong bones. You can get them from dairy products, such as milk and cheese.

Even if you are getting enough of these minerals in your foods, your body will not be able to use them properly unless you also have enough of certain vitamins, especially vitamin D. This vitamin helps the body absorb calcium and phosphorus from food in the intestines. It is formed by cells in the skin when they are exposed to the sun.

Athletes usually develop much stronger and heavier bones than people who do not get much exercise. If you had to stay in bed for a long time, your bones would become thinner and weaker. When no stress is put on bones, calcium salts are carried away by the blood, and the bones themselves actually become smaller. (When you got up again, you might be in some danger of breaking bones until they had a chance to grow strong again.) This kind of bone thinning can be a real problem for hospital patients, who spend a lot of time lying in bed. They need to do therapy to keep their muscles and bones moving so they don't wither away.

Wasting Away in Space

Astronauts who go on long space missions may "waste away" under weightless or low-gravity conditions. Without the pull of gravity, minerals are dissolved out of the bones and lost from the body. When astronauts go on long flights or stay in a space station in orbit, there is little or no gravity pulling on them. So their bones become weak, and the fine balance of minerals in their blood may be upset. The astronauts on early space missions were found to lose about 0.14 ounce (4 g) of calcium per month spent in space. Special exercises during spaceflights help keep astronauts' bodies in good condition and keep their bones from wasting away.

chapter four

Human babies look basically like little adults. As children grow, they get bigger and bigger each day, until one day, they are as big as their parents. But as they grow, they change mostly in size, not in shape. They already look much like their mothers and fathers. They usually have two eyes, two ears, a nose, a mouth, two arms, and two legs, just as their parents do. Many other animals grow gradually, just like us. A kitten looks like a little cat. A puppy looks like a little dog. A baby bird, with its beak and wings, looks very much like its parents.

Some babies in the animal kingdom do not look at all like their parents. If you saw a wormlike caterpillar, you might not think that it would grow up to be a butterfly, with large, brightly colored wings. A tadpole with just a big head and a long, swishing tail does not look at all like its tailless, hopping frog parents.

For a tadpole to turn into a frog or a caterpillar to turn into a butterfly, an amazing series of changes must take place. The baby must become almost a new animal before it can be like its parents. This change is called metamorphosis. (This term comes from Greek words meaning "a change in form.")

All living things have to go through some kind of changes as they grow and develop from a single cell to their complex adult form. The next two chapters describe the various kinds of development animals go through, ranging from the simple sponge to the more complex development of mammals.

The Animal Kingdom

The animal kingdom includes an extremely diverse group of organisms. There are more than one million different species (kinds of living things) on Earth. A small fraction of them—about forty-five thousand—are vertebrates (animals with backbones). But most of the members of the animal kingdom are invertebrates (animals without backbones). The largest group of invertebrates is the arthropods (meaning "jointed legs"). Arthropods include insects, crustaceans (such as crabs and lobsters), and arachnids (such as spiders and scorpions).

The Simple Life

Sponges are among the simplest animals. They have very little differentiation of the cells that make up their bodies. They

Giant barrel sponges, shown here in a cluster, are the simplest of animals.

have been called a republic of cells—that is, a group of cells living together and cooperating to get food and reproduce. Interestingly, if a sponge is forced through a wire mesh screen, breaking its body into individual cells, these cells can survive on their own. But if left in the same container, they soon find one another and join to re-form the sponge's body. No other animals can do this because their body cells are too differentiated and dependent on one another to survive alone.

When Life Gets More Complex

The hydra is a little more complex than the sponge. This pond-dwelling creature belongs to a group of animals that includes jellyfish, sea anemones, and corals.

The hydra's body shape is called a polyp—a long, slender vase with threadlike tentacles around the mouth, opening at the top. The tentacles are armed with poisonous stingers that attack tiny water

The Many-Headed Monster

The hydra got its name from Greek mythology, in which Hercules battled the Hydra, a many-headed monster. When Hercules cut off one of its heads, it grew two new ones. Much like its namesake, if a hydra is cut into pieces, each one will grow into a new animal.

creatures on which the hydra feeds. It lives its entire life as a polyp. It spends most of its time attached by its base to the bottom of the pond, a rock, or a stalk of a plant.

It is not uncommon to find a hydra with what looks like a baby hydra growing out of its side. Most hydras give birth to their young by budding. First, a little bump appears on the parent hydra's side. It is formed by mitosis of cells of the parent. At first they divide again and again without separating from the parent's body. The bump grows larger as cell division continues. Soon small tentacles begin to sprout, as genes for differentiation are turned on. The baby grows as a smaller copy of its parent. Finally, it breaks away and drifts to settle down and live a life of its own.

This hydra has reproduced by budding. The young hydra has grown out of the parent hydra's side and will continue to get larger until it breaks free to live on its own.

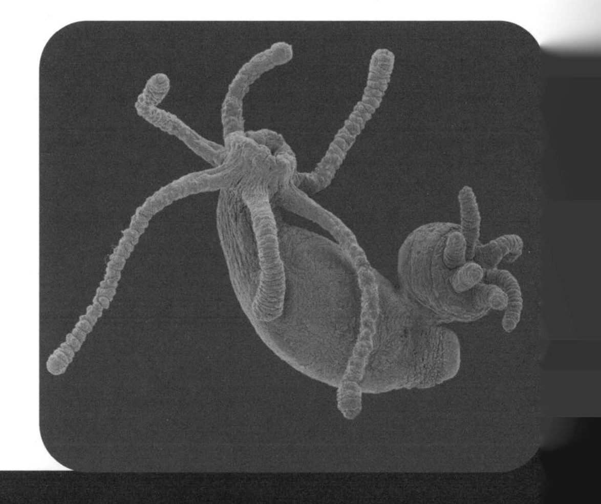

You are probably more familiar with the hydra's cousin, the jellyfish. You may have seen them in the ocean or washed up on the shore. Unlike the hydra, adult jellyfish have a dome-shaped body with poisonous tentacles dangling downward. Jellyfish don't actually swim. They are carried by ocean currents and drift or float through the waters, as their tentacles zap prey along the way.

The Snake-Haired Monster

Medusa comes from the name of the monster in Greek mythology that had snakes for hair (which look a lot like the tentacles of jellyfish). Scientists often call jellyfish medusas since they are not true fish. They are also called jellies or sea jellies.

While a hydra remains a polyp its entire life, a newborn jellyfish goes through a kind of metamorphosis to reach its adult form. When it's time to reproduce, the male jellyfish releases sperm cells into the water and the female releases egg cells. When a sperm joins with an egg, fertilization occurs, and soon a new life emerges in the form of a tiny swimming larva.

The larva has a pear-shaped body and doesn't look anything like its parents. After swimming around for a while, the larva attaches itself to a rock or some other object at the bottom of the sea and develops into a polyp. Then the polyp starts to divide, forming buds. This new colony of polyps looks like a stack of saucers. Each "saucer" in the stack breaks off from the colony and becomes a medusa, a free-floating jellyfish.

Strange Little Animals

Adult sea squirts look no more complicated than sponges or hydras. But looks can be deceiving. Sea squirts start life as highly developed swimming larvae, with features that are more like vertebrates than invertebrates.

Sea squirt larvae look like little tadpoles, much like those that turn into frogs—but these live in the ocean, not in ponds and lakes. The little "tadpoles" of the ocean swim freely, as their long tails send them gliding through the water. Eventually, though, their tails start to wag less and less, and soon they stop moving altogether. Slowly, the little creatures sink to the ocean bottom. Soon each larva attaches itself to a rock on the ocean bottom using three special suckers on the front of its head. There it stays, upside down, firmly fastened to the bottom. And then it starts to change.

The little animal's tail begins to disappear. It becomes shorter and shorter as the creature's body grows plumper and plumper. Its body twists about into the shape of the letter *U*, with what were its two ends pointing upward.

Even stranger changes are taking place inside the body. The little eyes, which helped the larva to swim toward the light near the surface of the water, are beginning to disappear. So are its tiny ears, which helped to keep its

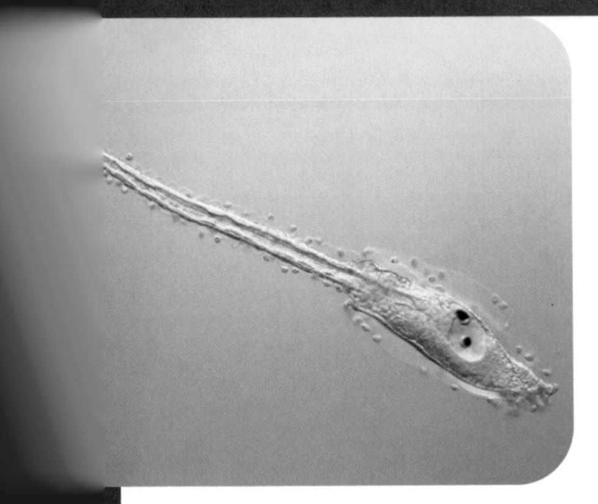

Sea squirt larvae like this one (left) *can swim around the ocean when they are young. Eventually, they settle to the bottom of the ocean and attach themselves to a rock or piece of coral, like the cluster of sea squirts above* (right). *They stay there for the rest of their lives.*

balance when it was swimming. The long cord of nerves, which passed down its back very much like our spinal cord, disappears too, leaving only a tiny brain. But its throat—covered with thousands of tiny openings, like a strainer—grows bigger and bigger until it almost fills the animal's body. Meanwhile, a thick, tough covering, called a tunic, forms on the outside of its body. (The sea squirt is also known as a tunicate.)

When these changes have all taken place, the little animal no longer resembles a vertebrate. It looks like a little round barrel, with two funnel-shaped "mouths," out of which may poke two long, hollow tubes. It sucks water in through one tube and squirts it out the other. (Hence, its common name *sea squirt.*)

An adult sea squirt leads a rather dull life compared to its younger self. It no longer swims freely in the ocean. Once it attaches itself to a rock or seaweed at the bottom of the sea, it never moves again. It grows larger and comes to look like a brightly colored rubber bag. It hardly seems like an animal at all.

All this strange little animal ever does is suck in water, strain out tiny bits of food through the strainers inside its body, and squirt the water out again. One day it will breed, releasing egg cells and sperm into the water. These cells will join with sperm and eggs from other sea squirts to make tiny "tadpoles" of the ocean. They will begin the amazing changes of metamorphosis all over again.

Stars of the Sea

You would surely recognize a starfish on the beach if you saw one, with its star-shaped body. But you probably would not recognize a baby starfish. It looks nothing like its parents. It does not have five arms like the points of a star. In fact, it doesn't have any arms at all.

Starfish begin their lives as tiny fertilized eggs floating in the water. These eggs start dividing in much the same way as human zygotes. The early development of the embryos is so similar to ours that eggs of sea urchins, a close

Did You Know?

Starfish aren't really fish. Long ago, people started calling them starfish because they looked like five-pointed stars and they lived in the water, like fish. Scientists call them sea stars, but the name *starfish* is still often used.

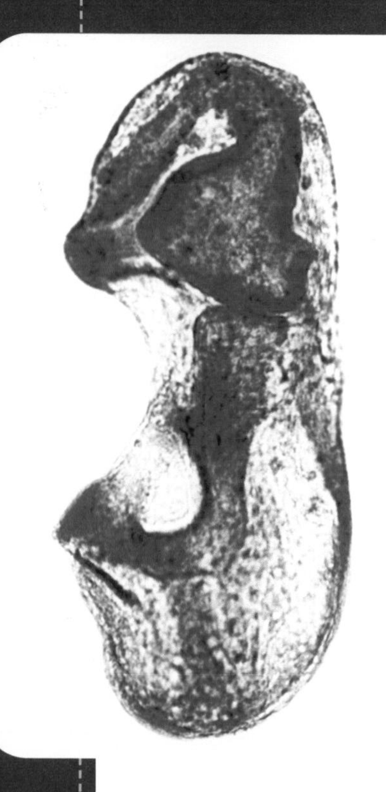

This larva of a starfish doesn't at all resemble the star shape it will have as an adult. The larva is able to swim through water, instead of crawling along the bottom of the sea as it does later in life.

relative of starfish, are used in school labs to study embryonic development. Starfish embryos do not develop into miniature humans, however, but rather into tiny swimming

A Starfish Surprise

Starfish can eat a whole bed of clams or oysters in a single night. For many years, whenever oyster gatherers caught a starfish, they chopped it into pieces and threw them back into the sea. They thought this was a good way to get rid of the pest. They didn't know that if a starfish loses an arm, it will grow a new one. In fact, the arm itself may grow into a whole new starfish! So they weren't killing the starfish at all. They were actually making more of them.

larvae. They look like knobby little kidney beans, with a small mouth and rows of tiny, hairlike cilia that wave back and forth and help them swim through the water. For a while they swim about and feed on little bits of green plant life.

After a few weeks, each little larva settles to the bottom and attaches itself to a rock with a special sucker that grows near its mouth. There it goes through some fascinating changes. Its mouth closes completely, and a new one opens in a different place. Five little arms appear where none were. Soon it becomes a five-pointed star, so small that it could fit on the tip of your thumb. It is very hungry and crawls away from the rock in search of tiny clams to eat. It will eat and grow with each passing day. Soon it will be a full-sized orange or brown or purple or green starfish, crawling about on the bottom of the sea.

Life in a Shell

The life cycle of a snail is much more direct than the ones we have discussed so far. A snail lays eggs—from twenty to a hundred or more. When the eggs are ready to hatch, they crack and little snails emerge—there is no in-between stage. They look like tiny versions of their parents. They even have their own little shells, but these shells are very thin and soft.

Did You Know?

Every snail is both a male and a female. When two snails mate, each one transfers sperm cells to the other. Later, each lays its own eggs.

A baby snail's first meal is its eggshell, which provides minerals for building its coiled shell. As the snail grows, the shell grows too and becomes harder and thicker. It is made of calcium, so it is hard like stone.

Not all animals' shells are rock hard like a snail's. Crabs, lobsters, and other crustaceans wear shells that are a bit more fragile and may break or crack. Their "shell" is actually their skeleton, which is on the outside of their body. It is called an exoskeleton. The exoskeleton is thicker in some places, such as the claws and back, acting like a suit of armor. But in other parts, such as the legs and joints, the exoskeleton is paper thin and can break off easily if pulled too hard.

Unlike a snail's shell, a crustacean's exoskeleton cannot grow. Once it hardens, it stays the same size. And so, like all the other arthropods, crustaceans must

A Crab's Tool Kit

Like other arthropods, a crab has a number of appendages attached to its body. Humans usually have just four appendages—two arms and two legs. Crabs have not only ten legs (including the two claws in front) but also a variety of other useful tools including mouthparts, feelers (antennae), and eyestalks (movable stems with eyes at the tips). Unlike ours, a crab's eyes and chewing tools are attached to the outside of its body.

This shore crab has shed its exoskeleton and will wait for its new shell to harden before it comes out of hiding.

shed their exoskeleton and form a new, larger one in order to grow. This is called molting. Crustaceans have to do this many times during their lives.

When a crustacean, such as a crab, is ready to molt, it looks for a safe hideaway, such as down in the sand, under leaves, or in a burrow. The old exoskeleton cracks open down the back, and the crab wriggles out of it. Underneath is the new exoskeleton, which is soft and moist. The crab is very vulnerable without a hard shell to protect it from predators. So it hides until the exoskeleton hardens. It takes a day or two for the new outer covering to harden. Then the crab can come out of hiding.

Crabs usually reproduce during times of molting. Several weeks to several months after the female lays her eggs, tiny swimming larvae emerge. These little creatures do not look

Borrowed Shells

Hermit crabs do not grow their own shells. Unlike other crabs, a hermit crab has a soft exoskeleton. It uses a snail shell as a suit of armor. As a hermit crab grows, its borrowed shell becomes too tight, and it needs to look for a bigger one. When it spots an empty snail shell, it will slip out of the old one and try the new one on for size. If the new shell fits, the crab will abandon the old, outgrown shell and walk away in its new armor.

like crabs at all. Instead, they look like tiny shrimp. Even at this early stage, the larvae are covered with a hard shell and have to molt to grow.

After molting several times, the larvae turn into megalops, which look like little crabs. (*Megalops* means "large eyes.") They develop the claws and the body parts of an adult crab. The megalops settle down on the ocean floor, where they finish their final stage of growing. After a few more molts, the megalops finally become small crabs. They will continue to molt as they grow bigger.

Creepy Crawlies

Thus far, we have explored the life cycles of a variety of organisms, but the most classic examples of metamorphosis can be seen in the insect family. More kinds of insects live on Earth than all other kinds of animals combined. Most insects go through what scientists call complete metamorphosis. Complete metamorphosis involves four stages: egg, larva, pupa, and adult. As an insect grows from the first stage to the last, its body undergoes a complete reorganization—the larva looks entirely different from the adult.

Giant Killer Creatures?

In scary science-fiction movies, giant insects bigger than humans threaten to kill every living thing they meet. In real life, though, insects could never grow that large. The exoskeleton to anchor their muscles would be so thick and heavy that its weight would crush them. The largest insects in the world are South American beetles that grow up to 6 inches (15 cm) long. Stick insects may grow to more than 15 inches (38 cm), but their long bodies are very thin. The owlet moth of tropical America has a wingspan of 18 inches (46 cm), but its body is much smaller.

Animals that live in the ocean can grow larger because the water helps to support them. The largest animal with an exoskeleton is the giant clam, tridacna. It lives in coral reefs in the tropics. It can grow as large as 4 feet (1.2 m) in diameter and weigh more than 500 pounds (227 kg).

Early Development

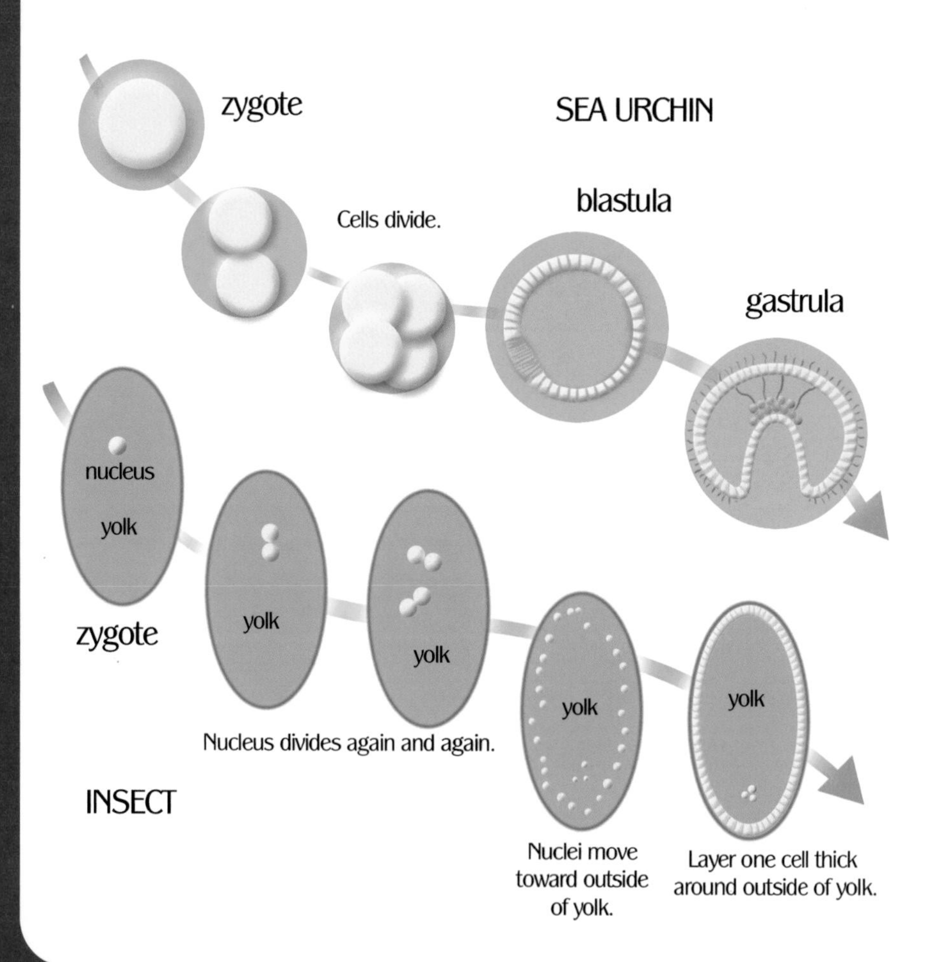

The early development of a sea urchin embryo is very similar to that of a human embryo. The fertilized egg divides to form a cluster of cells, then a hollow ball (blastula), and then a gastrula with several layers of cells. In insects, early development follows a different pattern. First, the nucleus of the fertilized egg divides many times to form about five thousand nuclei. These nuclei move apart toward the outside of the yolk, where they form a layer that's one cell thick.

The best-known example of complete metamorphosis is a caterpillar transforming into a butterfly or moth. A butterfly or moth goes through four stages of life. It starts as an egg, which develops somewhat differently than human or sea urchin eggs. An insect egg contains a lot of stored food (yolk), which will be used by the developing embryo. Instead of the whole egg dividing, at first just the nucleus (containing its DNA) divides—and divides again, twelve or thirteen times.

About five thousand daughter nuclei are formed and move to the outside of the yolk. They form a layer one cell thick around the outside of the yolk. On one side of the egg, cells from this layer grow larger, divide, and differentiate, forming the insect's body. Cells on the other side of the egg develop into thin membranes that form a covering for the embryo.

The young butterfly or moth that hatches from the egg is a larva called a caterpillar. The caterpillar has a long, wormlike body, with three pairs of legs and some tiny prolegs that cling, like Velcro, to leaves and twigs. A caterpillar is like a little

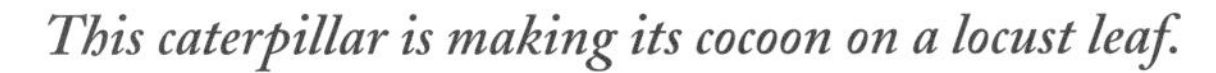

This caterpillar is making its cocoon on a locust leaf.

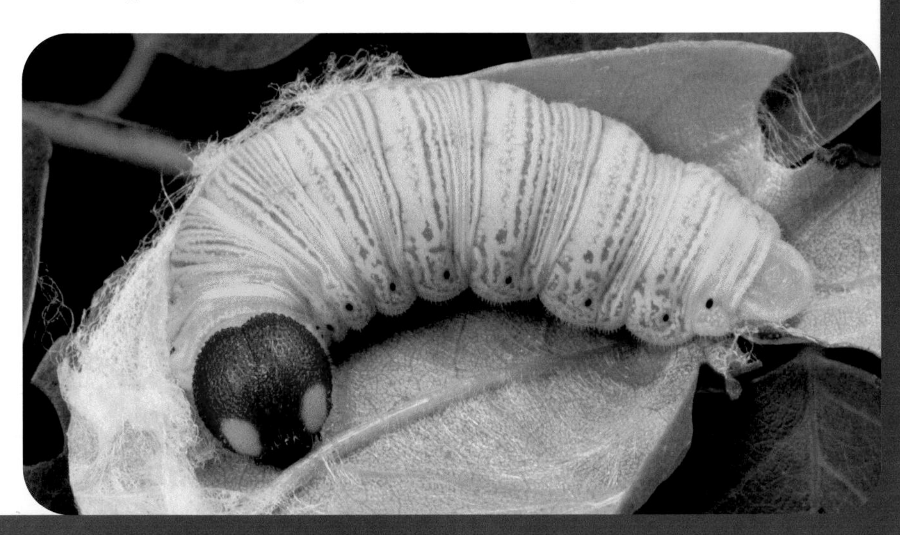

eating machine, gobbling up plant leaves to store energy for the big change. As the caterpillar grows, its body becomes too fat for its exoskeleton, which cannot stretch or grow. So the caterpillar must molt to become larger. Underneath is a new soft layer. It quickly hardens after the caterpillar wriggles out of its old exoskeleton. A caterpillar molts several times until it is fully grown.

The caterpillar attaches itself to a twig or branch or a burrow underground and forms a pupa. Some moth pupae are covered with a silk cocoon that the larva has spun around itself for protection.

Chemical Controls

What makes a caterpillar turn into a pupa and then develop into an adult? An insect's life plan is "written" in its genes, spelled out in a chemical alphabet. Messages from the genes are carried by hormones, just as they are in humans. Molting hormones cause an insect larva to shed its exoskeleton, permitting it to grow larger. Another kind of insect hormone, called juvenile hormone, keeps it in the larval form. When it is time to turn into an adult, the insect's genes stop making juvenile hormone. Then the next molt will produce a pupa instead of a bigger larva.

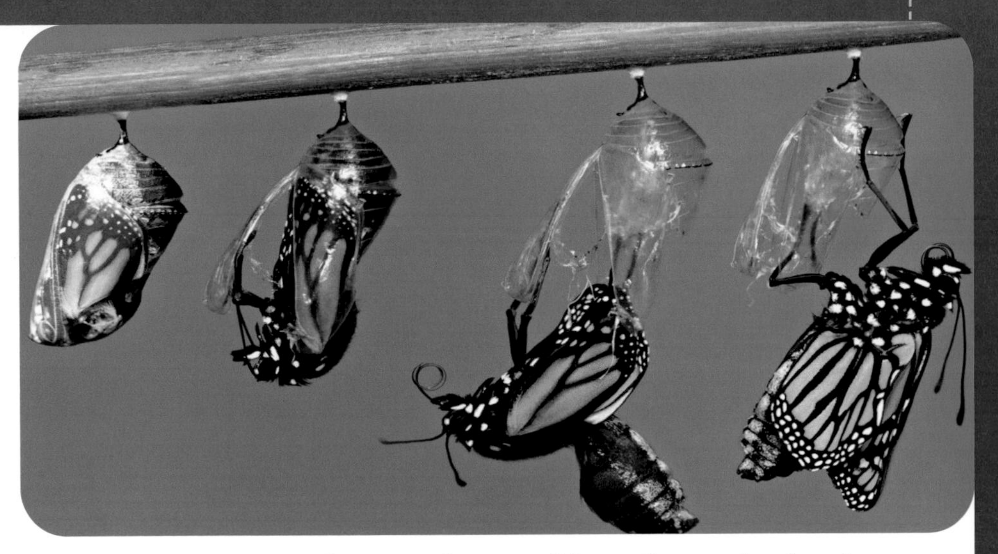

This image shows the stages of a monarch butterfly emerging from its chrysalis.

A butterfly pupa, called a chrysalis, is often green or multicolored and is not covered with a cocoon.

Inside the pupa's exoskeleton, everything is changing. The caterpillar's body turns into a soupy liquid, and then the adult form begins to develop. The body is much smaller and has three main parts: a head, a thorax (the middle section), and an abdomen. When it is time to emerge, the insect will molt once again and push its way out into the world. It spreads its wings to dry and is soon a butterfly or moth. While the caterpillar larva had munching mouthparts, most butterflies and moths have a strawlike mouth called a proboscis that is used for sucking nectar from flowers. All these changes take place within about ten days. But some butterflies stay in the chrysalis form over the winter and emerge in the spring.

Some kinds of insects, such as crickets, grasshoppers, and cockroaches, go through an incomplete metamorphosis. This process involves just three stages of development: egg, nymph,

Life Cycles

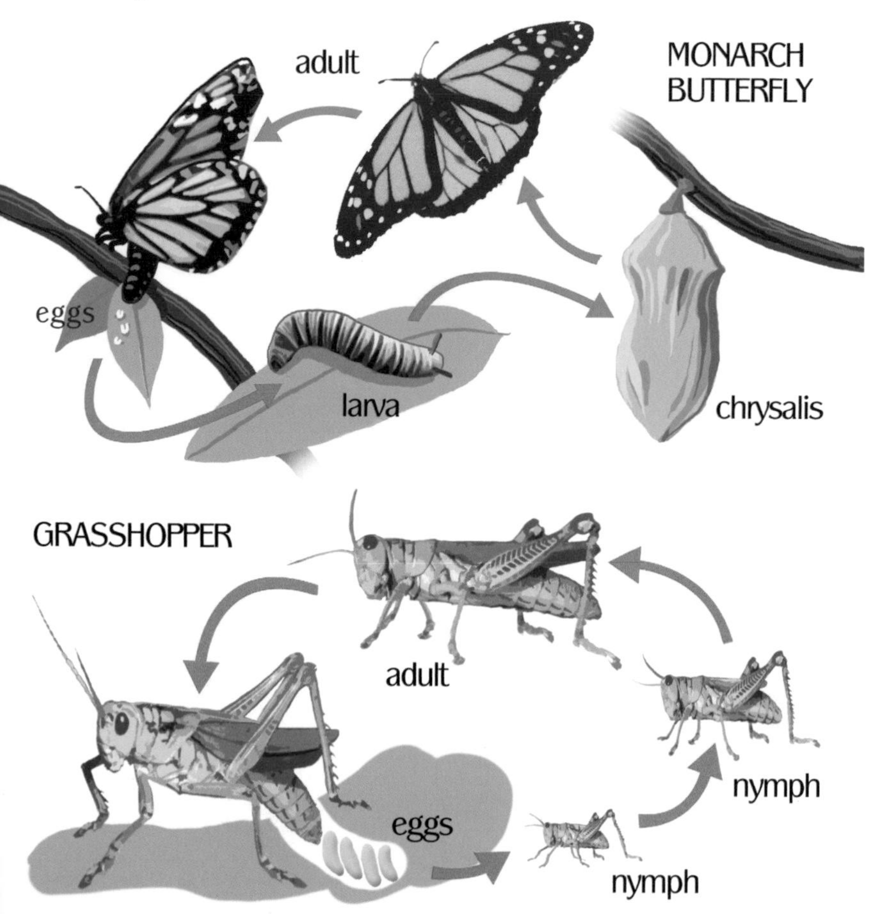

Butterflies, like many other insects, go through complete metamorphosis. An egg hatches into a larva, which then forms a pupa (chrysalis). Finally, the winged adult emerges. Grasshopper metamorphosis is incomplete. The egg hatches into a nymph, which looks somewhat like a miniature adult. The nymph molts and forms a larger nymph, which has longer wings and more segments in its body. Finally, the largest nymph molts and an adult emerges.

and adult. When the egg of a grasshopper hatches, for example, a tiny version of an adult grasshopper emerges. There is no larval stage. The young grasshopper, called a nymph, does not yet have wings, just stubs. It is also lighter colored than the adults. Like other insects a grasshopper has a tough exoskeleton and must molt to grow. It will molt many times until it becomes adult size and grows wings.

Super Organisms

Many organisms have specialized cells that work together to keep the animal alive and functional. But certain kinds of insects, including ants, bees, wasps, and termites, take the development of a species one step further. These superorganisms form specialized groups that work together in a colony. An ant colony, for example, is made up of hundreds, thousands, or even millions of ants. These ants must perform certain jobs to help ensure the survival of the colony. Some ants gather food, some dig tunnels, some are soldiers that fight to defend the colony, and some feed the queen (the only ant that can lay eggs) and take care of the young ants. These ants don't choose their jobs. They are born knowing exactly what they have to do.

chapter five

Animals with Backbones

Humans are related to all the living things on our planet—from sponges to snails, starfish, honeybees, and even bacteria and plants. But all those species are rather distant cousins on our family tree. Actually, modern humans are much more closely related to fish, the first major group of vertebrates that appeared on Earth. The early development of fish embryos is similar to how human embryos develop.

Fish are water animals. They eat, sleep, and produce their young in oceans, lakes, ponds, and streams. Their streamlined bodies are covered with a layer of scales, thin bony plates that overlap like the shingles on a roof. Fanlike fins send them gliding through the water. They do not need to breathe air. They take in oxygen from the water that flows over their gills.

Many fish reproduce by spawning—males and females release their sex cells into the water much like sea squirts and starfish. The sperm and egg cells join to make new fish. Fish that spawn usually produce large numbers of eggs. Female herring, for example, lay from twenty thousand to as many as two

Fun Fish

A 1996 issue of the scientific journal *Development* had a special feature: the upper right corners of 230 pages formed a flip book. The diagrams showed the development of a zebra fish embryo from two cells to hatching. These little fish, with their bold stripes, have long been aquarium favorites. Developmental biologists are raising them to learn more about vertebrate development. They are easy to raise and lay a lot of eggs—in the water, where scientists can watch them grow.

The embryos are transparent, so researchers can see what is going on inside them. Zebra fish embryos develop in just five days, and the fish can have young of their own when they are only three months old. Treating the fish with small amounts of chemicals produces hundreds of interesting changes in the genes controlling development. "It's an incredibly fun time now to work with zebra fish," said Oregon researcher Donald A. Kane in the December 7, 1996, issue of *Science News*.

hundred thousand eggs. Many fish eggs are eaten by sea animals before they can develop completely.

Some fish mate, and the young develop inside the mother's body. In some species, such as mollies, guppies, stingrays, and some sharks, the eggs contain all the nourishment the embryos need. When they hatch, the young fish swim out of their mother's body. In other species that bear live young, such as hammerhead sharks, ocean perch, and some aquarium fish, the embryos attach to the inside of the mother's body and receive nourishment from her.

A Double Life

What could be more interesting than an animal that leads a double life? We're talking about amphibians. *Amphibian* comes from a Greek word that means "double life." Amphibians do have a kind of double life—they can live on land and in the water. Frogs and toads make up the largest group of amphibians, which also include salamanders and newts.

Frogs and toads are well known for their dramatic changes in development. They go through a complete metamorphosis, in which they start life as fishlike, plant-eating tadpoles and eventually turn into completely different animals, ones that hop around and eat bugs.

When frogs hatch from eggs, they come out as wriggling little creatures, with big, round heads and long, flat tails. The baby frogs do not look at all like their parents. Instead, they look like little fish. They are called tadpoles or pollywogs.

For a long while, the little tadpoles do not seem to be changing much at all. They swim about in the pond, feeding on small bits of green plant life and growing larger and larger. They are like fish in many ways. Tadpoles do not have legs. And they must stay in the water all the time because they do not have lungs for breathing air. Instead, tadpoles breathe through gills on the sides of their heads, just as fish do.

As the tadpole starts to change, two little knobs, or buds, appear near the back of the tadpole's head—one on each side, close to where the head joins the tail. With each passing day, the buds grow larger. Soon they begin to look like legs, at first short and stubby, then longer with webbed toes. These are hind legs, and the tadpole is a little odd-looking at this stage. It has a big head, a tail, and a pair of long hind legs, but no forelegs can be seen. Then in a week or two, two more buds appear, farther forward. Within days they grow into perfectly formed forelegs.

Some of these tadpoles are just starting to grow hind legs as they develop into adult frogs.

The Link

Scientists believe that amphibians developed millions of years ago as a link between fish and reptiles. Ancient amphibians had heads and tails like fish, but they had lungs for breathing air and legs for moving around on land. They returned to the water to lay their eggs. As generations went by, some amphibians lost their tails and developed into frogs and toads. A few lost their legs and became snakelike. Millions of years later, some of the amphibians evolved into reptiles. The reptiles could lay their eggs on land and spend their lives out of the water.

The little tadpole is looking more and more like a frog. Interesting changes are going on inside its body, too. It is growing a pair of lungs so that it can breathe air, and its gills are starting to disappear. Teeth are growing in its mouth. And its stomach and intestines are changing. A tadpole has very long intestines to digest the green plant life it eats. After a while, though, the intestines begin to get shorter. The tadpole is preparing for a new way of life. As a frog it will not eat plants at all. It will feed on insects and other animals, which are easier to digest.

For a time, the newly developed creature is in a kind of in-between stage—it is no longer a tadpole, but it is not yet a frog. It can no longer digest the plant life it once ate, but it is not yet able to catch prey. And so the tadpole stops eating. But it does not starve. It has a source of food stored away in its body—its tail! The almost-frog begins to digest its own tail and use it for energy. Each day the tail grows shorter, until at last it disappears. Finally, the change is complete.

Salamanders look like little lizards, but they are actually amphibians with tails. They also have to go through several stages of metamorphosis, but they differ somewhat from frogs and toads. Sometime after the female lays her eggs in the water, young salamanders, called larvae, emerge from the eggs.

Salamander larvae are different from the tadpoles of frogs. The larvae have obvious, feathery external gills (the gills that tadpoles use to breathe are internal, like those of a fish). The front legs develop before the back ones (the opposite of frogs). Also, tadpoles feed completely on algae and other vegetation,

A Larva That Never Grows Up

Salamanders called axolotls, which live in Mexico, do not finish their metamorphosis. They keep their feathery gills and do not grow lungs. They do not need organs for breathing air because they spend their entire lives in the water. Axolotls do have all the genes for normal salamander development, though. If the hormone thyroxine is added to the water where an axolotl is living, it will grow lungs, lose its gills, and be able to live on dry land.

while salamander larvae are carnivorous (meat eaters) and eat all the tiny water insects they can find. Eventually, the growing young salamanders move onto land, where they continue to develop into the adult form. During metamorphosis, frogs lose their tails, but salamanders keep theirs.

Roaming Reptiles

Unlike amphibians, reptiles can live entirely on land. They have lungs for breathing air and are covered with a protective layer of scales, made of a tough substance somewhat like our fingernails. The scaly skin keeps their bodies from drying out. Although some reptiles, such as sea turtles, sea snakes, alligators, and crocodiles, spend much of their time in the water, others, such as lizards and snakes, can live in very dry areas—even deserts.

Most reptiles lay eggs, although some give birth to live young. Reptile eggs do not have to develop and hatch in the water. They can be laid under rocks or in holes in the ground. Inside the thick, leathery protective shell, the reptile egg contains a food supply (yolk) for the developing young and is wrapped in a baglike covering. The babies that develop inside these cozy sacs hatch as smaller versions of their parents. They are ready to run about and eat, even shortly after birth. (A newly hatched poisonous snake can bite right after hatching.)

A snake grows throughout its entire life. But a

Snakes with Legs?

Snakes are legless reptiles. But it wasn't always that way. Millions of years ago, snakes evolved from lizards. Like lizards these ancient snakes had four legs. Because they did a lot of burrowing, the snakes had very little need for their limbs. Some of their offspring had shorter limbs or even none at all. This did not hurt their chances of survival. In fact, the snakes that lost their limbs had a more streamlined shape and were very successful in their burrowing life. The legless snakes multiplied and began to outnumber the lizardlike snakes. Thus, over many, many generations, snakes developed into their present-day appearance and a new way of moving around. Modern pythons and boa constrictors still have the remains of their ancestors' leg bones as little nubs under their skin.

snake's tough, scale-covered skin cannot stretch as the snake grows. So the snake has to molt (shed its skin) regularly to grow. (Lizards molt too.)

Free as a Bird

Birds and reptiles actually have some things in common. Birds, too, have scaly skin. Scientists believe that's because birds evolved from reptiles. But bird eggs are not leathery like those of a reptile. Instead, a bird's eggs are protected by a hard shell.

One or both parents keep the eggs warm with their own bodies while the chicks develop. The baby chicks have a lot of growing to do inside the eggs until they are ready to enter the outside world. Even after they hatch, they still have quite a bit of developing to do before they are ready to go out on their own.

Growing Back Body Parts

All living things have the ability to regenerate—replace destroyed, defective, or damaged tissue with new growth. This ability varies greatly from species to species, though. In humans and other complex animals, regeneration is limited. People can heal their wounds, produce new blood cells, and replace damaged tissue in broken bones and some organs. However, animals such as lobsters, crabs, salamanders, and lizards can replace entire limbs. Lizards will sometimes detach all or part of their tail as a defense from predators. The tail breaks off cleanly, distracting the predator long enough for the lizard to safely scamper away. Later, the lizard will grow a new tail to replace the missing one. However, the new tail may be a different color, size, or texture. Flatworms and hydras, on the other hand, can grow entire new bodies from a small fragment.

This mother northern goshawk feeds her chicks. The parents are responsible for feeding their chicks many times a day.

Some chicks are born blind, featherless, and helpless. They need to be fed many times a day by their parents. But the food they eat in the first days of life has to be mashed up. So the parents eat the food first, then regurgitate (spit up) the partly digested food into their chicks' open mouths. As they get older, the young birds will be able to handle solid food brought to them by their parents.

Gradually, the chicks develop feathers and can see. The parents continue to take care of their young and teach them after they become independent, for up to two years. Some kinds of behavior, however, such as flying, are not learned—a bird already knows how to do them.

Some chicks, such as those of chickens, turkeys, ducks, and geese, have soft, fluffy down (feathers) and can see well when they hatch. These chicks can run around and peck at bits of food soon after hatching.

Some Birds Can't Fly

All birds have wings, but that doesn't mean that all birds can fly. The largest living birds are the North African ostriches. They can grow as tall as 9 feet (2.7 m) and weigh as much as 345 pounds (157 kg). Their wings are small compared to their huge bodies and far too weak to lift them up into the air. Other birds that don't fly are emus and kiwis. Like ostriches their wings are small and weak. Penguins do not fly through the air, but they can "fly" very well in the ocean. Their wings are strong flippers that propel them through the water. You might think that chickens and turkeys can't fly. Actually, they can fly for short distances, although they spend most of their time on the ground.

All birds have to molt. Before baby birds can fly, they lose some of their down as the adult feathers grow in. In many species the first real molt takes place about three or four months later. Then every year thereafter, the adult bird molts again, around the end of the breeding season. The old, worn-out feathers are replaced with new ones. This doesn't happen all at once. A few feathers fall out and are soon replaced by new ones. This process continues again and again, until all the feathers have been replaced. The bird is never bald during molting, and it never loses so many flight feathers that it cannot fly.

How Mammals Grow

The most highly developed animals are mammals. Mammals are warm-blooded vertebrates whose females feed their young with milk produced in their mammary glands (that's how mammals got their name). Many mammals give birth to live young that are small and helpless. The offspring depend on their mother's milk to grow and develop. They spend a long time with the adults, as they learn the skills they need to survive.

Many mammals live on land, but some, such as whales, dolphins, seals, and manatees, live in water. Most mammals have hair or fur that acts as insulation to keep body heat from escaping. Even whales and dolphins, which are nearly hairless as adults, have hair at some stage in their development. (In adults, a

They're All the Same but Different

All vertebrates look like one another in certain stages of their early development. A mammal embryo goes through fishlike and reptilelike stages during its development before birth. For example, fish, turtles, chickens, mice, and humans all develop fishlike tails and gills during the embryo stage. However, only fish continue to develop gills, and only fish, turtles, and mice keep the tails. (A chicken's tail is made of feathers and is not really a part of its body.)

Development in placental mammals varies a great deal from animal to animal. These red fox babies are born blind and helpless.

thin layer of fatty blubber under the skin acts as insulation.) Bats are the only mammals that fly—flying squirrels actually glide.

Three main types of mammals exist: placental, monotreme, and marsupial. Most mammals are placental mammals. The young of these mammals develop inside the mother's body, receiving nourishment through the placenta. This group includes humans, as well as mice, dogs, cats, bears, monkeys, elephants, and many others.

Development varies a great deal from species to species. Some animals are born at a very immature stage, before they have finished developing. Newborn young of many rodents (such as mice and hamsters) and carnivores (such as bears and foxes) are completely helpless. Their eyes are closed, they can't hear, and they can barely move. But newborn horses and sheep can walk within a few minutes. Within an hour, they can run with their mother.

Monotremes are very unusual mammals. This group includes the duck-billed platypus—an animal with features

The duck-billed platypus is one of the few kinds of mammals that lays eggs to reproduce. The platypus is also an unusual mammal because it can use venomous poison to protect itself.

that resemble a bird, a reptile, and a mammal. It has a bill and webbed feet like a duck. It has poison glands and lays leathery eggs like a reptile. It has fur, is warm-blooded, and feeds its young with milk as a mammal does. (The mother monotreme does not have nipples, however. Instead, the young lap up milk that flows from pores in the mother's skin.) This group also includes two species of echidnas, or spiny anteaters. They also have fur, lay eggs, and feed milk to their young. Monotremes are found only in Australia, New Guinea, and Tasmania.

Marsupials give birth to tiny, undeveloped young.

Young marsupials like this baby kangaroo may stay with their mothers until they are about a year old. The mother's pouch provides nourishment and safety.

A Fish with a Pouch

Marsupials are not the only animals with pouches. The sea horse is actually a fish with a pouch. But this is the only thing these two animals have in common. For one thing, it is the male sea horse, not the female, that has a pouch—and it is the father that becomes pregnant. The baby sea horses grow and develop inside the father's pouch. The pregnancy can take as long as six weeks. When the male finally gives birth, tiny sea horses burst out of his pouch. Then they can live on their own.

Some are born as little as eight days after fertilization. Each tiny baby, about the size of a kidney bean, crawls up the mother's belly into a pouch. It clamps its mouth onto one of the mother's mammary nipples and stays in the pouch, safe and protected, for months until it is ready to come out. Some well-known marsupials are kangaroos and koalas. Most marsupials live in Australia and on nearby islands. The only marsupials found in the rest of world are opossums.

chapter six

How Plants Grow

What is the largest living creature on Earth? Blue whales may grow up to 100 feet (30 m) long and weigh as much as 200 tons (181 metric tons)! But the giant sequoias and redwood trees of California grow far larger than that. A sequoia named the General Sherman Tree is more than 270 feet (82 m) tall and measures more than 100 feet (30 m) around the base of its trunk. Its weight is estimated at about 6,000 tons (5,400 metric tons).

Some redwoods grow to more than 350 feet (107 m) tall, although their trunks are not as broad. These giants belong to the plant kingdom. The other 400,000 or so species of plants on this planet include a great variety, from giant trees to tiny pond-living duckweeds, just 0.02 inch (0.5 mm) across.

Despite their diversity, plants do share a number of features. Like all living things, plants are made up of cells. Just as when you grow and develop, cells are also important in the growth and development of plants. Growing a new plant leaf or a longer tree

The giant sequoia named the General Sherman Tree is considered one of the world's largest organisms. The General Sherman is found at Sequoia National Park in California and is approximately 2,200 years old.

branch involves cell growth and cell division.

Just like animals, plants need raw materials—food—to grow. But they don't eat. Oak trees, dandelions, and tiny algae in a pond can do something that no animal can do: they can use sunlight energy to make their own food. They do this by photosynthesis.

Sunlight energy (*photo*) is used to put simple chemicals together to form more complex ones (*synthesis*). Plants need water and carbon dioxide for photosynthesis. Using the energy of sunlight, they turn these raw materials into sugars, starches, and other carbohydrates.

Green plants get their color from chlorophyll, a green pigment in the leaves that absorbs sunlight energy that can be

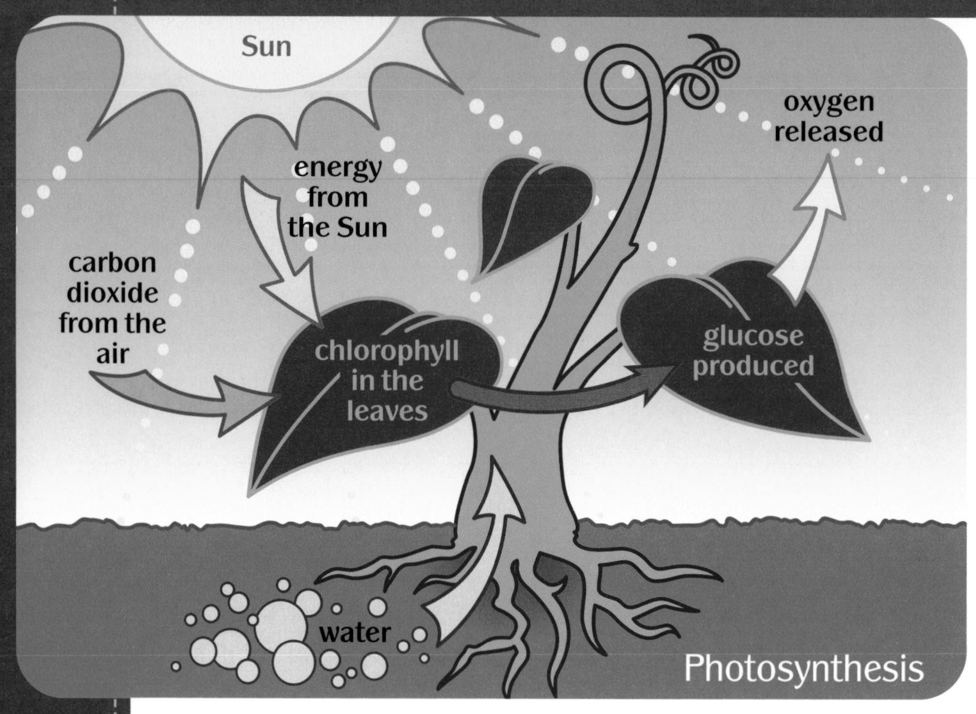

Photosynthesis is the process by which green plants make their food. The process begins when sunlight interacts with the chlorophyll in the plant's leaves and light energy is stored as chemical energy in the chloroplasts. Water and carbon dioxide also enter the leaf. The carbon dioxide combines with water, using the energy stored in the chloroplasts, to produce sugar. The sugar is then transported to other parts of the plant and either stored, used right away for energy, or used to make other food substances.

used to power the chemical reactions of photosynthesis. Part of the sunlight energy is converted to chemical energy, which is stored in the plant's tissues. Photosynthesis takes place both in the water and on land. The need for sunlight explains why many water plants live in shallow waters rather than the deep sea.

It Just Keeps Growing

American gardeners got their first glimpse of the Japanese plant kudzu in 1876. They loved its large leaves and sweet-smelling flowers. Soon kudzu was being planted in gardens and fields throughout the southern United States. It looked pretty, provided food for animals, and helped keep the soil from washing away. The only problem was that it grew too fast. Kudzu vines can grow up to 1 foot (0.3 m) a day in the summer and 60 feet (18 m) in a year. They climb trees and telephone poles and can kill entire forests by preventing the trees from getting sunlight. People are still trying to figure out how to stop kudzu from spreading. It already covers more than 7,000,000 acres (2,800,000 hectares) of land and has been called the vine that ate the South.

The Japanese plant kudzu has taken over this area in Mississippi, spreading over the trees and everything in its path. The U.S. government once paid people to plant kudzu. Now it is considered a weed and costs $500 million each year in lost croplands and control measures.

The Simplest Plants

The simplest plants have a rather small number of cells. Each cell can take in raw materials directly from the air or water in which they live. Over many millions of years, some plants got larger and more complex. These plants needed a "plumbing system" of pipes and tubes to transfer water and food materials from one part to another. Scientists call such plants vascular plants. The tubes that run through them are somewhat like the blood vessels that carry materials through the bodies of most animals.

Meanwhile, as plants grew more complex, so did their lifestyle. In some cases, growth and development involve changes just as dramatic as those in the metamorphosis of frogs or butterflies.

This tree fern found in Costa Rica (left) *started out as a small plant early in its life cycle* (right). *All ferns go through both stages as they develop.*

Patterns of Life

If you lay a fern frond down on a sheet of paper and tap it gently, then lift it carefully, you will find a pattern of dots on the paper. They form an outline of the fern frond. These dots are fern spores, cells that can reproduce to form new plants. They are found on the underside of the frond. A breeze or passing animal may shake the frond, dropping spores onto the ground. There they sprout into new plants that do not look at all like fronds. New fronds will later sprout when these tiny plants reproduce.

Ferns, for example, live two lives. Two very different-looking forms take turns in a fern's life cycle. Each one lives, grows, and changes into the other form when it reproduces. The feathery leaves (fronds) of a fern give rise to small, hardly noticeable plants. Those tiny plants reproduce, in turn, forming the familiar kind of fern with feathery fronds. Scientists call this kind of life cycle alternation of generations.

Mosses are smaller, simpler plants that do not have systems of tubes to transfer materials. But they, too, go through an alternation of generations. Mosses are tiny, delicate plants that live in moist soil. They reproduce to form even smaller plants, which live as parasites on their parents. The small plants are attached to the larger parents and get all their food materials from the parent plants. They form spores that grow into the larger mosses.

Seed Plants

Most plants are even more complex than mosses or ferns. Instead of alternating generations, the sex cells of the parent plants join to form seeds, which combine the parents' DNA (containing their hereditary instructions). Seeds may seem lifeless. They may go for long periods without moving, growing, or photosynthesizing. But add some moisture, and suddenly they "wake up" and sprout. Then they grow into a whole new plant that can live on its own.

Seeds start out as fertilized eggs, which develop in an organ of the flower called an ovary. (A fruit is really a ripened ovary.)

Each seed contains an embryo, a tiny plant with a miniature root and a shoot that includes a stem and one or more folded-up leaves. The seed also contains

Seed Development

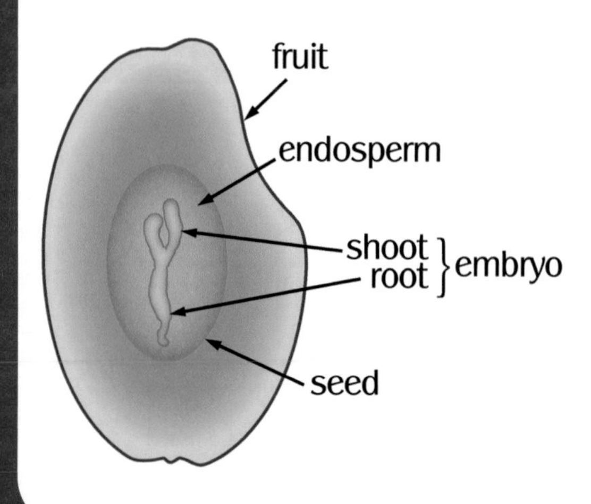

The seed of a flowering plant is usually enclosed in a fruit, which may be carried away or eaten by animals. The hard seed inside the fruit includes stored food (endosperm) and a tiny embryo with a root and a shoot (stem and leaves). The root and shoot will grow much larger when the seed sprouts.

These corn seedlings have sprouted up out of the dirt into corn plants. The roots have also grown and extended farther into the ground.

a supply of stored food. The sprouting plant can use this food until it grows big enough to start making its own food through photosynthesis.

Plants have specialized growing areas called meristems in the root and shoot tips. These areas contain cells that have not differentiated and thus could grow into any kind of specialized plant cell. Cell division in the meristem helps the plant grow taller and extends its root system.

Another area of meristem, called the cambium, contains cells that are partly differentiated. Cell division of the cambium helps the plant grow wider and thicker around the stem and form the pipes and tubes that transport fluids within the plant. Some plants grow a single large root, called a taproot, which may reach deep into the soil. The roots of other plants grow many branches that spread over a wide area.

Plant stems vary too. Some are short and delicate. Others grow tall and may be strengthened by woody materials, forming sturdy tree trunks. Growing taller or growing branches that

Growing like a Beanstalk

Have you ever taken apart the two halves of a lima bean? Hidden inside is a tiny embryo plant. If you plant a bean seed in soil and water it, the embryo sprouts. Its tiny root grows longer and may curve around so that its tip is pointing downward. Its little stem grows longer and turns so that its leaves are pointing up. After a time, it is long enough to poke out of the soil. The two fleshy halves of the seed, called the cotyledons, provide the food for this growth. When the seedling straightens up out of the soil, the cotyledons are at the top.

Soon the cotyledons spread out, turn green, and start to produce more food by photosynthesis. With the strong head start provided by the cotyledons, the bean plant grows quickly. When true leaves have formed and they have grown big enough to make food for the growing plant, the cotyledons shrivel and fall off. The plant branches and forms more leaves. Soon it is producing more food than it needs for new growth. Then it can start forming flowers, which produce the fruits—long pods containing seeds for a new generation.

These strawberry plants have sent out runners from the mother plant. The runners will go into the ground at different places and grow new roots and leaves, letting the plant reproduce.

spread out sideways helps plants compete for sunlight to use for photosynthesis. Some plants can also use new growth to reproduce. Strawberries send out runners—stems that dip down to the ground at various spots and form new roots and leaves. If the stems connecting these plantlets to the main plant are cut, they can grow into new, individual plants.

Seed plants belong to two main groups: the gymnosperms (evergreens such as pine trees and firs, often used as Christmas trees) and the angiosperms (flowering plants such as daisies, tulips, cherry trees, and dogwood trees). *Gymnosperm* means "naked seeds." The seeds in a pinecone, for example, are bare except for a thin covering. *Angiosperm* means "covered seeds." The seeds of these

Seeds that Travel

Not all flowering plants produce fruits that are good to eat. The "fluff" on a dandelion that has gone to seed is the part of its fruit that helps the seeds travel to new locations, riding on the winds. Cocklebur fruits have spiky little hooks that stick to the fur of passing animals or people's clothing. Some fruits, such as witch hazel pods, even explode, shooting their seeds out and away from the mother plant, so the seeds are scattered widely.

flowering plants are usually found inside a fruit, formed from a part of the flower.

Some fruits, such as cherries or peaches, contain a single seed. Others, such as apples and blueberries, contain many seeds. Some foods that we think of as vegetables are really fruits—for example, string beans and green peppers. They develop from ripened ovaries, just as cherries, oranges, and grapes do.

Just as in animals, hormones help control and coordinate growth and development in plants. Several kinds of hormones are involved. Auxins are formed in the tips of a plant's roots and shoots. From there they move through the tissues to other parts of the plant. *Auxin* comes from a Greek word meaning "to increase." Small amounts of these hormones help cells become longer and then divide. As a result, the growing roots and shoots become longer and thicker. (Oddly enough, although small amounts of auxin help growth in the roots, larger concentrations stop roots from growing. In the shoots, any amount of auxin promotes growth.) Auxins help develop the tubes that carry materials through a plant's stems and leaves. They prompt tree buds to open, forming new growth, and they help fruits to develop.

The hormone cytokinin (named from Greek words meaning "cell" and "to move") is produced in roots, young fruits, and seeds. It makes cells divide faster and helps keep plants from growing old and dying.

The hormone gibberellin was first discovered by a researcher in Japan who was studying "foolish seedling" disease in rice plants. A fungus living on the plants produces a chemical that causes them to grow fast, with long, pale stems that break easily. Later, it was found that similar chemicals are

One of a Kind

The oldest seed to sprout and grow into a plant was a two-thousand-year-old date palm seed from Masada, in ancient Israel. It was found on an archaeological dig in the 1970s, and researchers sprouted it thirty years later, in 2005. By then no more trees of its kind were growing anywhere in the world. The great forests of date palms up to 80 feet (24 m) tall that existed in biblical times were gone, replaced by desert.

formed by all the seed plants. Like auxins, gibberellins prompt cells to grow longer and to divide. These hormones make plants grow taller and also help seeds to sprout. Dwarf plants have a change in their genes that prevents them from making normal amounts of gibberellins.

Ethylene is the only plant hormone that is a gas. It prevents growth and helps fruits to ripen and drop off the plant. Plants produce ethylene when they have too little water or minerals, damage by pests or plant-eating animals, and other kinds of stress. By saving

Plants Move by Growing

Unlike animals, plants do not have muscles to move their bodies. But they do move, although they cannot get up and walk around. A sunflower turns its blossoms toward the sun and keeps turning through the day as the sun moves across the sky. Bean plants spread their leaves during the day and close them tightly at night. A morning glory vine will twine around anything that touches it. These and other plant movements are produced by growth.

If a seed is planted so that its embryo is lying on its side, the sprouting shoot will bend upward and the root will bend downward. That happens because the energy that would be needed for growth and development, the plants may be able to survive until conditions get better.

The plant hormones work together to control the development of a plant in various stages of its life cycle. Gibberellins, auxins, and cytokinins help seeds to sprout and grow. The growth of a young seedling is very rapid for a while, as it forms new leaves and makes food by photosynthesis. Then growth slows down as some of the food is used to form flowers and fruits.

Annual plants stop growing and start to die after they have produced their fruits. Their seeds are dormant—they are alive,

---

auxin flows from the tips of the root and shoot, but gravity causes it to pool on the underside of the sprout. The underside of the shoot will grow faster, causing it to bend upward. In the root, the larger amount of auxin on the underside prevents growth, so the upper side grows faster and the root bends downward. Soon the growing seedling is right side up.

A plant turns its leaves or flowers toward the sun because sunlight destroys auxin. So the side of the stem facing the light stops growing, but some auxin still is in the shaded side of the stem. That side continues to grow, and the stem bends over.

but they do not seem to be doing anything. They can stay dormant through the winter or through a dry season. When the next growing season brings warm temperatures and rain, the seeds will sprout, grow, and bear fruits of their own.

Perennial plants continue to grow year after year. Most of the growth takes place in the spring. Growth slows down during the hot, dry periods of the

These cross sections of three tree trunks each tell a different life story. The numbers of rings show the age of the trees, and the darker and lighter areas indicate periods of faster and slower growth.

summer. As the weather turns cold in the fall and winter, some perennial plants may appear to die. But their roots in the ground are still alive, in a dormant state.

Trees also become dormant in the winter, and some of them lose all their leaves. (Evergreens such as pines and firs keep their needlelike leaves through the winter, but they do not grow.) The rings that can be seen in the woody trunk when a tree is cut down are records of the tree's growth during each year of its life. Darker and lighter areas mark periods of slow or rapid growth. Scientists can tell the age of a tree by counting its rings. They can also learn how weather conditions varied from year to year by studying the patterns of the annual rings.

chapter seven

Future Development

What if humans could grow a new arm or leg to replace a missing or damaged one, the way lizards or lobsters can? That sounds like science fiction. But it may actually be closer to reality than you think, thanks to advances in medical technology. These days, researchers have developed laboratory techniques for growing tissues that can replace bone, cartilage, blood vessels, and skin. In fact, this technology was put to the test in the mid-1990s when doctors performed experimental surgery on a twelve-year-old boy named Sean McCormack.

Sean was born without cartilage or bone under the skin on the left side of his chest, due to a rare condition known as Poland's Syndrome. Without cartilage or bone protecting it, Sean's heart could actually be seen beating in his chest. Doctors had talked about implanting an artificial plate when Sean was twenty-one, after he had finished growing.

Sean McCormick, shown here at age sixteen, had a laboratory-grown cartilage plate surgically placed in his chest when he was twelve.

By twelve, however, Sean was getting impatient. He was a star pitcher for his Little League baseball team, and without protection for his heart, he was in danger every time he played ball. Sean's doctor suggested that the family talk with a group of scientists and surgeons at Children's Hospital in Boston who were doing research on growing human body parts artificially.

Surgeons at the Boston hospital wanted to implant a laboratory-grown cartilage plate into Sean's chest. The operation would be experimental and risky. But Sean's family decided to go ahead with it. The procedure was a success, and the new cartilage cells took hold in Sean's body. Within a year, his chest appeared normal and was able to grow along with him. Four years after the operation, at sixteen, Sean commented about the plate in a July 1998 article in *Business Week*, "It's pretty cool. It looks like something I was born with."

Growing Body Parts

Researchers used some of Sean's own cartilage cells to grow his new chest. These cells were placed on a framework of plasticlike material, adding special hormones to make them grow. The framework was biodegradable, so that after the cells grew over it and formed cartilage and other tissues, it could eventually be broken down and absorbed by the body.

Researchers have been making a lot of progress in tissue engineering—growing new tissues, organs, and body parts to replace damaged parts and to treat diseases. They have used cartilage grown in the lab to make artificial replacements for sections of the intestines. They have even built an entire trachea (windpipe) that was successfully transplanted into a sheep. Tissue engineers have also built new heart valves, parts of the liver, and complete bladders.

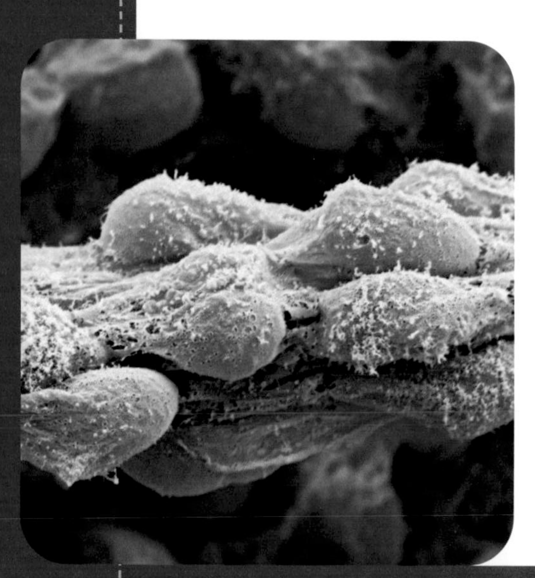

This scanning electron micrograph gives a three-dimensional view of human embryonic stem cells. These stem cells have grown and formed into a horizontal strand.

What Are Stem Cells?

The first cells in a developing embryo have not yet differentiated. Each one could develop into any kind of cell or tissue in the body—skin cells, heart cells, liver cells—or even grow into a whole new organism. Scientists call these immature cells stem cells. As the embryo develops further, some of its cells start to differentiate. They are then committed to forming particular kinds of cells. By the time a baby is born, it still has many stem cells. Its umbilical cord also contains blood rich in stem cells. Even in an adult, a tiny fraction of the cells in the bone marrow, liver, spleen, and other tissues are stem cells. The body uses them to form new tissue to repair damage or replace cells that have died.

Tissue engineers can also make replacement body parts from stem cells. These cells can grow and differentiate to form new tissues. Stem cells transplanted into a damaged heart grow into new heart muscle tissue. Other stem cells have been grown into working islet cells. Islet cells are cells in the pancreas that produce the hormone insulin. Some people with diabetes cannot make enough insulin in their own pancreas. Islet cells are helping these people control their blood sugar level without having to take insulin injections.

All this progress has been made possible by research on how growth and development take place. Researchers have learned that growth factors—hormones and other chemicals—

help to direct the processes, turning genes in the cells on and off at just the right times to form tissues and organs. The right mix of growth factors can even cause stem cells to develop into particular kinds of tissues. Stem cells from bone marrow, for example, normally produce various kinds of blood cells. But researchers have grown bone marrow stem cells into

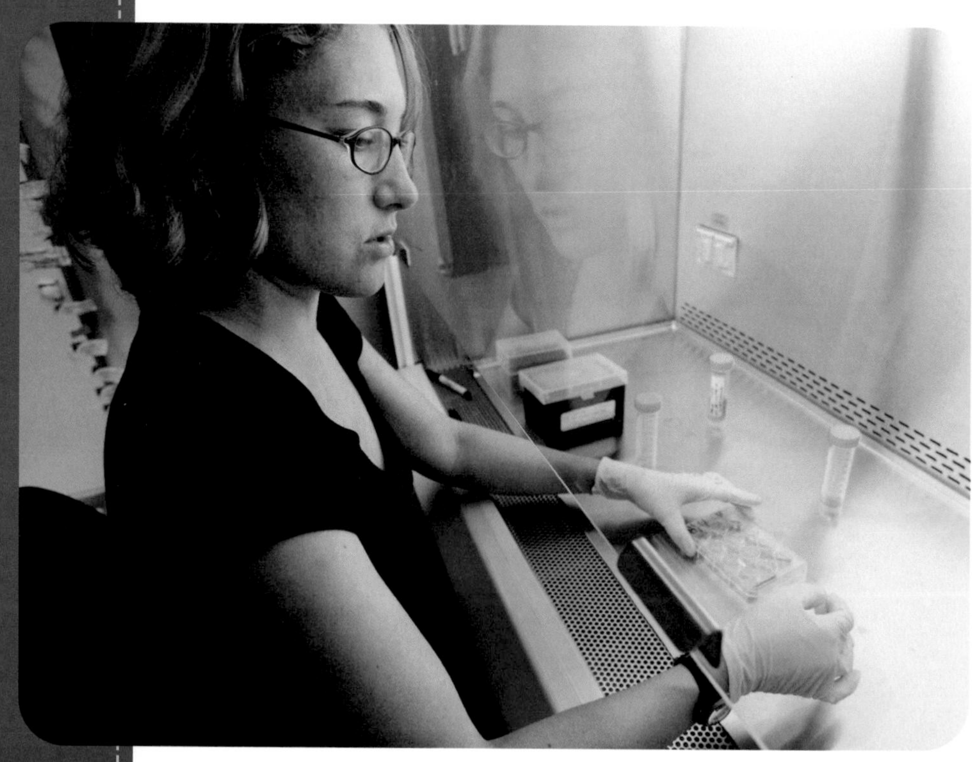

This graduate student works with stem cell cultures in a lab at the University of California-Irvine in 2006. Stem cell research has led scientists to learn more about the genes that control development.

muscle and nerve cells, using suitable growth factors.

Researchers have also been learning more about the genes that control development. They have found genes that help direct the work of other genes that form proteins and build body tissues. With this growing knowledge, they hope someday to be able to stimulate regeneration, making the body itself grow replacement parts when an organ is damaged.

They also hope to correct errors in genes and even to design new working genes that will guide the development of new organisms. For example, fixing the genes that cause blood diseases such as sickle-cell anemia would be a cure for these inherited diseases. Tinkering with the on-off switches of growth genes and the hormones that control them could allow dwarfs to grow taller.

Further in the future, researchers might try more dramatic changes. Human cells contain genes for forming gills, for example. Normally, these genes are turned off at an early stage of development and the gills in an early human embryo disappear. Turning the genes that direct the formation of fishlike gills back on might give people the ability to breathe underwater. Adding genes for producing chlorophyll to the set of genes that form skin might allow us to make some of our own food, just as plants do.

The Evo-Devo Revolution

After scientists discovered that genes are made of DNA, they began to learn a lot about how cells work and how organisms develop. In the 1980s researchers discovered a number of genes controlling the development of the fruit fly, from egg to larva to adult. Meanwhile, other researchers were gradually learning to "read" the DNA of animals from worms to humans. Comparing the genes controlling fruit fly development to the DNA of other species led to a surprising discovery.

Most biologists had believed that the various body parts—such as eyes, limbs, and hearts—of humans and other vertebrates evolved separately and independently from those in insects and other invertebrates. But it turned out that frogs, mice, butterflies, and even humans have sets of genes controlling development that are almost exactly the same as those of the fruit fly. And they do the same things.

Learning from Nature's Mistakes

Genes controlling development were first discovered by studying the DNA of "genetic monsters"—animals that developed abnormally. For example, mice, fruit flies, and even humans are sometimes born without eyes. All of them turned out to be missing a particular gene. Mutant fruit flies without wings, limbs, or antennae have also been found. Researchers have even found some fruit flies with a leg instead of one of the antennae, or a leg that grew out of the back instead of a wing. All of these have added to our knowledge of what genes do.

This bullfrog developed abnormally from a mutant gene and ended up with six legs, two on one side and four on the other. All of the legs actually work.

A gene in mice, for example, causes eyes to form. Fruit flies have their own version of this gene. If the mouse gene is inserted into a developing fly embryo, in the part that will form a wing in the adult fly, the fly will grow an extra eye on its wing! This eye will be a compound eye, made up of hundreds of tiny lenses. So the mouse gene caused a *fruit fly eye* to form, even though mice, like humans, have eyes with a single lens. The gene somehow told the embryo, "Make an eye here." Other genes in the insect said, "Make it a fruit fly eye."

It seems that all the animals have a standard tool kit of master genes for forming body parts and building them into a working organism. Studies of fruit flies revealed that eight genes control the development of body regions: three for the head, three for the thorax, and two for the abdomen (the hind region). These genes are all together, in the same order as the parts of the fly's body—from head to "tail." (The parts of an animal embryo's body develop in this order too.)

Most of the DNA in each gene is quite different from the DNA in the other seven body-forming genes. That is not surprising, since the head, thorax, and abdomen, and the things that are attached to them, are quite different. Curiously, though, all eight genes have one short piece of DNA that is the same. Scientists named these genes *Hox,* from homeobox (*homeo* is a Greek word for "same").

Later, it was found that mice also have *Hox* genes, which control the development of parts of the body. *Hox* genes have been found in many other animal species as well, including humans. And they all have almost exactly the same short piece of DNA—even animals as different as fruit flies, mice, elephants, and humans. In fact, the same piece of DNA is also found in bacteria.

What do *Hox* genes do? In bacteria they make proteins that bind to DNA. These proteins cover genes and prevent them from working. If conditions inside the cell change, the binding proteins on certain genes may come off, leaving the genes free to work. Thus, the DNA-binding proteins act as on-off switches for the genes. The *Hox* genes in animals also produce on-off switches. As an embryo develops, different combinations of "control genes" produce binding proteins, turning the genes for making body parts on and off as development proceeds.

All the discoveries in genetics and development have led to an active new area of biology called evo-devo. It is the study of how new life-forms **evo**lve

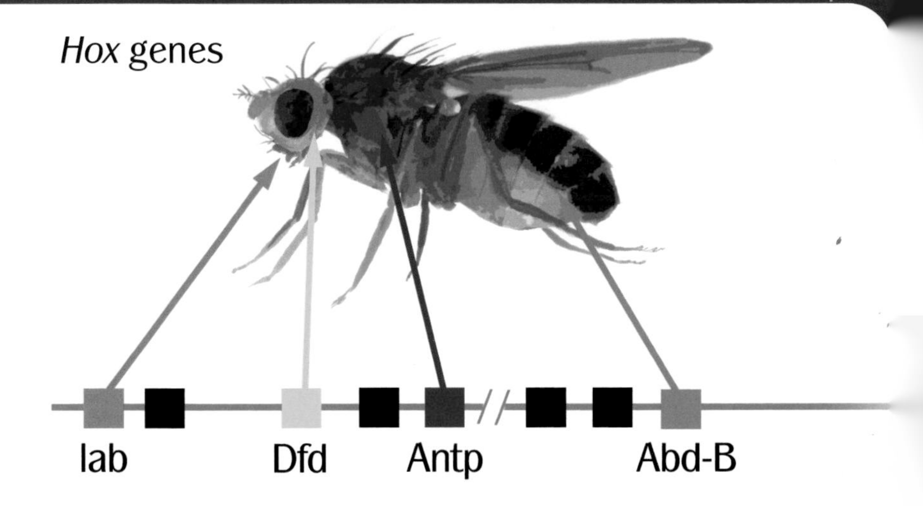

Some of the fruit fly's genes control the development of specific parts of the body. Extra copies of one of these genes would make that body part longer. The letters below the colored boxes are abbreviated names of the genes involved.

through changes in the **dev**el**o**pment of individuals. (The boldfaced letters show where the term *evo-devo* came from.) Some general ideas have already emerged.

An animal's body, for example, might be thought of as a series of segments, or blocks—like a chain of snap-together blocks or the cars in a railroad train. When cells copy their DNA and divide, errors can occur. Sometimes an extra copy of a gene may accidentally be made. Another gene may be skipped over and lost. Still another might be changed, switched for a different gene, or shifted to a different part of the sequence. These events are rare, but over the long history of life, there have been plenty of opportunities for them to occur.

An extra copy of a *Hox* gene could make the body longer. If this gene controls the formation of a body segment with a pair of legs, the offspring may have an extra pair of legs. A missing *Hox* gene would result in a shorter body with fewer legs. A change in *Hox* genes might produce legs where wings should have formed.

The animal kingdom has many examples of such changes in body form over long periods of time. The millipede, for instance, is an ancient member of the group of arthropods. It has a long body with many similar-looking segments. Most segments have two pairs of walking legs. Crustaceans are also arthropods but have ten legs. Some are claws used for grasping. Spiders and scorpions have only eight legs. Insects have only six legs but may also have one or two pairs of wings.

During evolution the number of body segments at first increases but then decreases. Meanwhile, some body parts may be changed as old body-forming genes are put to new uses. In general, the mutations that succeed in the struggle for survival and are passed on to future generations produce only small changes. Major changes are more likely to make an organism less fit to survive and pass on its genes. A long series of small changes, accumulating over extremely long spans of time, gradually produced the marvelous diversity of life on our planet. Scientists working in the new evo-devo field are tracing the story of how all this occurred.

Thus, studies of growth and development are not just providing new insights into the past history of life on Earth. They are also showing ways to improve our own bodies and those of the other life-forms that share our planet.

Glossary

abdomen: in arthropods, the last section of the body

alternation of generations: a pattern of development in which two very different forms take turns in the life cycle

amniotic sac: the membrane that surrounds the developing embryo, which floats in amniotic fluid

amphibian: an organism that can live both in the water and on land; a group including frogs and salamanders

angiosperms: flowering plants in which the seeds are covered by a fruit

annual plant: a plant that lives for only one growing season

arthropod: an invertebrate with jointed limbs, a segmented body, and an exoskeleton made of chitin

auxin: plant hormone that promotes the growth of roots and shoots

bacteria: single-celled organisms that do not have chlorophyll and reproduce by mitosis

benign tumor: a tumor that may grow very large but does not invade nearby tissues

binary fission: a form of reproduction consisting of the splitting of a cell into two daughter cells

budding: a process by which offspring develop as an outgrowth of a parent

cambium: specialized growing area in plants that helps them grow wider and thicker around the stems

cell: the basic unit of living things

cell division: the process by which one cell divides into two

chlorophyll: a green pigment used by plants to absorb light energy; used for photosynthesis

chrysalis: the hard-shelled pupa of a butterfly

cilia: hairlike structures on the outside of a cell; used for movement

cocoon: the silk covering that an insect larva forms around itself before entering its pupal stage

cotyledon: the part of a seed that contains stored food. In some sprouted seeds, the cotyledons grow into temporary leaves that make food by photosynthesis until the true leaves can take over.

cytokinin: plant hormone, produced by the roots, that promotes tissue growth and budding

differentiation: a process by which cells become different and specialized

DNA (deoxyribonucleic acid): the chemical basis of hereditary traits

dormant: inactive; in a resting state

egg: a female sex cell

embryo: the early stages of development of an animal in the womb or egg; in plants a tiny plant with a miniature root and a shoot that includes a stem and one or more folded-up leaves developed inside a seed

endocrine: pertaining to hormones and the structures that produce them

ethylene: plant hormone that prevents growth and helps fruits ripen and drop off the plant

evo-devo: evolutionary developmental biology; a field that studies the links between evolution and the development of different organisms

evolution: the theory of successive development of life-forms through changes in the genetic material

exoskeleton: a skeleton that forms an animal's outer covering

fallopian tubes: the pair of tubes that carry the egg from the ovary to the uterus

fertilization: the joining of an egg and a sperm, producing a new individual with traits inherited from its mother and father

fetus: a human embryo after the ninth week of development

fronds: the leaves of ferns, consisting of hundreds of tiny leaflets

genes: chemical units that determine hereditary traits passed on from one generation of cells or organisms to the next

gibberellin: growth hormone produced in fungi and plants

growth factors: hormones and other chemicals that promote the growth, organization, and maintenance of tissues and organs

gymnosperms: seed-producing plants that do not have flowers, including evergreens such as pine trees, spruce, and firs

hormones: chemicals that help control and regulate the body's activities

***Hox* gene:** a gene that contains a homeobox (a short section of DNA that is almost identical in all genes that contain it)

human growth hormone: a chemical secreted by the pituitary gland that promotes growth

invertebrate: an animal without a backbone

juvenile hormone: an insect hormone that prevents metamorphosis into the adult form

larva: an immature form of an animal, which usually looks quite different from the adults of its species

malignant tumor: a cancerous tumor, which grows uncontrollably and invades nearby tissues

mammal: a vertebrate with fur that produces milk to feed its young

marsupial: a mammal that does not have a placenta and gives birth to immature young, who complete their development in a pouch on the mother's abdomen; includes kangaroos and opossums

medusa: the scientific name of a jellyfish, an umbrella-shaped animal with the body opening at the bottom

meristem: specialized growing areas in plants that help them grow taller and extend their root system; located in the root and shoot tips

metamorphosis: a series of major changes in form during an animal's development

mitosis: the process of cell division producing daughter cells with exactly the same DNA instructions as in the mother cell

molting: shedding the outer covering of the body

monotreme: an egg-laying mammal, which includes the duck-billed platypus and the echidnas of Australia and New Guinea

mutation: a change in DNA that can lead to a change in structure, function, or both

nucleus: the control center of a cell, containing its DNA

nymph: the young of an insect that undergoes incomplete metamorphosis

organ: a grouping of tissues in a specific structure, such as a heart or kidney

perennial plant: a plant with a life cycle that lasts for more than two years

photosynthesis: the process of turning carbon dioxide and water into sugars and starches, using the energy from sunlight

pituitary gland: an endocrine gland deep inside the brain that secretes hormones that control the production of hormones by the other endocrine glands

placenta: a structure in mammals that provides nourishment for a growing embryo and removes its waste products by an exchange with the mother's bloodstream

placental mammals: mammals that have placentas for the nourishment and development of their offspring

polyp: a vase-shaped form of a cnidarian (a type of invertebrate), with the body opening at the top, surrounded by tentacles

proboscis: long, tubular mouthpart of insects

protein: a biochemical made of long chains of units called amino acids, linked together. Most cell and body structures are built from proteins. Other proteins help to control chemical reactions.

puberty: the period of rapid growth and changes in the body as the sex organs mature and become capable of reproduction; secondary sex characteristics develop

pupa: a stage in an insect's life cycle, during which major reorganization of internal organs occurs within a protective outer case

regeneration: the replacement of destroyed, defective, or damaged tissue with new growths

reptile: a cold-blooded vertebrate with an outer covering of scales or bony plates

shoot: the part of a seed embryo that will grow into the aboveground parts of a plant

spawning: reproduction by release of sperm and egg cells into water, where they are fertilized outside the parents' bodies

species: a kind of living thing; a group of organisms that look similar and can breed with one another

sperm: a male sex cell

spore: a single cell that can grow into a plant

stem cells: immature cells that are not differentiated and can develop into various kinds of cells and tissues

teratoma: a tumor made up of several kinds of tissues. It may contain hair, teeth, or other body parts. Teratomas can be either benign or malignant.

thorax: the part of the body of an insect between the head and the abdomen

tissue: an organized group of cells with a similar structure and a common function

tissue engineering: laboratory techniques used for growing new tissues, organs, and body parts to replace damaged parts and to treat diseases

trimesters: the three parts of a woman's pregnancy, each three months long

tumor: a solid mass formed from a buildup of cells

umbilical cord: a cord of tissue, containing blood vessels, that connects the developing embryo to its mother's uterus

uterus: the womb; the organ in which a fertilized egg grows into a baby

vascular plants: plants with an arrangement of tubes, running from the roots through the stems to the leaves, that conducts water and food materials

vertebrate: an animal with a backbone

wart: a tumor that grows from a single skin cell infected by a virus

zygote: a new cell formed by the joining of an egg and a sperm

Bibliography

Arnst, Catherine, and John Carey. "Biotech Bodies." *Business Week*, July 27, 1998. http://www.businessweek.com/1998/30/b3588001.htm (August 30, 2006).

Biology-Online.org. "Tutorials: Growth Patterns." *Biology-Online.org*, January 1, 2000. http://www.biology-online.org/3/10_growth_patterns.htm (July 18, 2005).

Canfisco. "Salmon Life Cycle." *Salmon Education Center*. 2006. http://www.goldseal.ca/wildsalmon/life_cycle.asp (August 29, 2006).

Carroll, Sean B. *Endless Forms Most Beautiful: The New Science of Evo Devo*. New York: Norton, 2005.

Dowshen, Steven. "Growth Problems." *Nemours Foundation*, October 2004. http://www.kidshealth.org/teen/diseases__conditions/growth/growth_hormone.html (July 29, 2005).

Dunn, Gary A. "Introduction to Insects." *Young Entomologists Society*. http://members.aol.com/YESedu/introbug.html (June 14, 2005).

Gilbert, Scott F. *Developmental Biology*. 6th ed. Sunderland, MA: Sinauer, 2000.

Goodman, Corey S., and Bridget C. Coughlin. "The Evolution of Evo-Devo Biology." *PNAS*, April 25, 2000, 4424–4425. http://www.pnas.org/cgi/content/full/97/9/4424 (July 21, 2005).

Hall, Brian K. "Evo-Devo Is the New Buzzword." *Scientific American*, April 2005. http://www.sciam.com/article.cfm?articleID=0005D708-2F7C-123B-AF7C83414B7F0000&sc=I100322 (July 21, 2005).

Hoagland, Mahlon, and Bert Dodson. *The Way Life Works*. New York: Times Books, 1995.

Homeier, Barbara P. "I'm Growing Up, but Am I Normal?" *Kids Health*, April 2005. http://www.kidshealth.org/kid/grow/body_stuff/growing_up_normal.html (July 29, 2005).

IUPUI Department of Biology. "Human Reproduction and Development." *IUPUI Department of Biology*, March 3, 2004. http://www.biology.iupui.edu/biocourses/N100/2k4ch39repronotes.html (July 28, 2005).

Kalman, Matthew. "Seed of Extinct Date Palm Sprouts after 2,000 Years." *San Francisco Chronicle*, June 12, 2005. http://www.sfgate.com/cgi-bin/article.cgi?f=/c/a/2005/06/12/MNGJND7G5T1.DTL&hw=extinct+date+palm&sn=001&sc=1000 (July 18, 2005).

Malley, C. "Preschooler Development." *Family Day Care Facts* series. Amherst: University of Massachusetts, 1991. National Network for Child Care (NNCC) reprint, 1995, 2002. http://www.nncc.org/Child.Dev/presch.dev.html (August 29, 2006).

Meyer, John R. "Insect Development: Embryogenesis." NC State University. February 17, 2006. http://www.cals.ncsu.edu/course/ent425/tutorial/embryogenesis.html (August 30, 2006).

Silverstein, Alvin. *Human Anatomy and Physiology*. New York: John Wiley & Sons, 1983.

Silverstein, Alvin, Virginia B. Silverstein, and Laura Silverstein Nunn. *Cells*. New York: Twenty-First Century Books, 2002.

———. *Puberty*. New York: Franklin Watts, 2000.

Smith, John Maynard. Shaping Life: *Genes, Embryos and Evolution*. New Haven, CT: Yale University Press, 1998.

Sullivan, Jim. "Plant and Animal Cells." *Cells Alive!* 2004. http://www.cellsalive.com/toc.htm (July 29, 2005).

Tanner, James M., and Gordon Rattray Taylor. *Life Science Library: Growth*. New York: Time, 1965.

ThinkQuest USA. "Crabs." *What Swims Beneath: Sea Life*. 2002. http://library.thinkquest.org/CR0215242/crabs.htm (July 29, 2005).

Travis, John. "Gone Fishing!" *Science News*, December 7, 1996. http://www.sciencenews.org/pages/sn_arch/12_7_96/bob1.htm (August 30, 2006).

Wikipedia."Metamorphosis." *Wikipedia*. September 1, 2006. http://en.wikipedia.org/wiki/Metamorphosis_(biology) (September 5, 2006).

For Further Information

Books

Balkwill, Frances R., and Mic Rolph. *Enjoy Your Cells.* Woodbury, NY: Cold Spring Harbor Laboratory Press, 2001.

———. *Gene Machines.* Woodbury, NY: Cold Spring Harbor Laboratory Press, 2002.

———. *Have a Nice DNA.* Woodbury, NY: Cold Spring Harbor Laboratory Press, 2002.

Fleisher, Paul. *Evolution.* Minneapolis: Twenty-First Century Books, 2006.

Fridell, Ron. *Genetic Engineering.* Minneapolis: Lerner Publications Company, 2006.

Ganeri, Anita. *Plant Life Cycles.* Chicago: Heinemann Library, 2005.

Goodman, Susan E. *Seeds, Stems, and Stamens: The Ways Plants Fit into Their World.* Minneapolis: Millbrook Press, 2001.

Johnson, Rebecca L. *Genetics.* Minneapolis: Twenty-First Century Books, 2006.

Kalman, Bobbie. *The Life Cycle of a Butterfly.* New York: Crabtree Publishing Company, 2002.

———. *The Life Cycle of a Frog.* New York: Crabtree Publishing Company, 2002.

Silverstein, Alvin, and Virginia Silverstein. *Metamorphosis: Nature's Magical Transformations.* Mineola, NY: Dover Publications, 2003.

Spilsbury, Louise, and Richard Spilsbury. *Plant Growth.* Chicago: Heinemann Library, 2003.

Stewart, Melissa. *Maggots, Grubs, and More: The Secret Lives of Young Insects.* Minneapolis: Millbrook Press, 2003.

Websites

The Great Plant Escape

http://www.urbanext.uiuc.edu/gpe/

This is a fun-filled site in which you become a "detective" assigned to uncover the mysteries of plants.

Lisa Smart Asks, "How Does My Body Grow?"

http://www.sickkids.on.ca/kids/ks_Growth.asp

This KIDScience site is filled with facts on how fast kids grow and why.

Plants

http://www.ed.gov/pubs/parents/Science/plants.html

This website includes instructions for some simple experiments on what plants need to grow and how they use sunlight to make food.

Sea Urchin Embryology

http://www.stanford.edu/group/Urchin/contents.html

Labs, information, links, and some great animations on sea urchin development can be found on this website.

Index

Photo Acknowledgments

The images in this book are used with permission of: Getty Images (© Joseph H. Bailey/National Geographic, p. 5; © Fritz Goro/Time & Life Pictures, p. 16; © Steve Allen/The Image Bank, p. 24; © Luca Trovato/Stone, p. 29; © Yellow Dog Productions/The Image Bank, p. 32; © Keystone/Hulton Archive, p. 35; © Heather Perry/National Geographic, p. 40; © Dr. Dennis Kunkel/Visuals Unlimited, p. 41; © Tim Laman/National Geographic, p. 44 (right); © George Grall/National Geographic, p. 53; © Gary Vestal/ Photographer's Choice, p. 55; © Kim Taylor & Jane Burton/Dorling Kindersley, p. 61; © Michael S. Quinton/ National Geographic, p. 67; © Jane Burton/Dorling Kindersley, p. 70; © Altrendo Panoramic/Altrendo, p. 75; © R H Productions/ Robert Harding World Imagery, p. 78 (left); © Hiroshi Higuchi/Photographer's Choice, p. 88; © Sandy Huffaker, p. 94); Visuals Unlimited (© DWSPL, p. 9; © Ralph Hutchings, p. 26; © Wim van Egmond, p. 44 (left); © Biodisc, p. 46; © William Jorgensen, p. 49; © Rob & Ann Simpson, p. 97); © Visuals Unlimited/ CORBIS, p. 11; © Laura Westlund/ Independent Picture Service, pp. 12, 76; Photo Researchers, Inc. (© Nigel Cattlin, p. 18; © Claude Edelmann, p. 21 (all); © Biophoto Associates, p. 78 (right); © Professor Miodrag Stojkovic, p. 92); Ron Miller, pp. 23, 53, 56, 80, 99; © SuperStock, Inc./SuperStock, p. 71; PhotoDisc Royalty Free by Getty Images, pp. 72, 81; © Buddy Mays/CORBIS, p. 77; © Dwight R. Kuhn, p. 83; © Brooks Kraft/CORBIS, p. 91.
Cover: © Paul Gilham/Reportage/Getty Images.

About the Authors

Dr. Alvin Silverstein is a former professor of biology and director of the Physician Assistant Program at the College of Staten Island of the City University of New York. Virginia B. Silverstein is a translator of Russian scientific literature.

The Silversteins' collaboration began with a biochemical research project at the University of Pennsylvania. Since then they have produced six children and more than two hundred published books that have received high acclaim for their clear, timely, and authoritative coverage of science and health topics.

Laura Silverstein Nunn, a graduate of Kean College, began helping with the research for her parents' books while she was in high school. Since joining the writing team, she has coauthored more than eighty books.